A Guide to
Endangered and Threatened Species
in Virginia

A GUIDE TO
ENDANGERED AND THREATENED SPECIES
IN VIRGINIA

Coordinated by

Karen Terwilliger
Nongame and Endangered Wildlife Program
Virginia Department of Game and Inland Fisheries

and

John R. Tate
Office of Plant and Pest Services
Virginia Department of Agriculture and Consumer Services

Abridgment by
Susan L. Woodward
Radford University

The McDonald & Woodward Publishing Company
Blacksburg, Virginia
1995

The McDonald & Woodward Publishing Company
P. O. Box 10308, Blacksburg, Virginia 24062-0308

A Guide to Endangered and Threatened Species in Virginia

© 1994 by Department of Game and Inland Fisheries, Commonwealth of Virginia

All rights reserved. First printing 1995.
Composition by Rowan Mountain, Inc., Blacksburg, Virginia.
Printed in Canada by DWFriesen.

01 00 99 98 97 96 95   10 9 8 7 6 5 4 3 2 1

Library of Congress Cataloging-in-Publication Data

A guide to endangered and threatened species in Virginia /
coordinated by Karen Terwilliger and John R. Tate ; abridgement by
Susan L. Woodward.
    p.   cm.
Abridgement and update of: Virginia's endangered species. 1991.
Includes index.
    ISBN 0–939923–31–9 (alk. paper)
    1. Endangered species—Virginia. 2. Nature conservation—Virginia.
I. Terwilliger, Karen, 1954–   . II. Tate, John R., 1946–   .
III. Woodward, Susan L., 1944 Jan. 20–   IV. Virginia's endangered
species.
QH76.5.V8G85   1995
574.5'29'09755—dc20   95–23668
                                                            CIP

A Guide to
Endangered and Threatened Species
in Virginia

Dedicated
to the memory of

Lewis M. Costello

member of the Virginia Department of Game and Inland Fisheries
Board from July 1987 until his untimely death on April 30, 1993.
His remarkable energy, insight, dedication, and commitment were
a true inspiration. He left us a legacy of caring and service which
will long be remembered.

Produced
with the generous assistance of

Turner Sculpture
Onley, Virginia

Additional funding for this project was provided by
Nongame and Endangered Wildlife Program
Virginia Department of Game and Inland Fisheries
Virginia Department of Agriculture and Consumer Services
United States Environmental Protection Agency

Contents

Preface

This book represents an abridgment and update of *Virginia's Endangered Species,* an award-winning volume produced by the various state agencies responsible for the conservation of Virginia's biological resources (Virginia Department of Game and Inland Fisheries, Virginia Department of Agriculture and Consumer Services, Virginia Department of Conservation and Recreation, and the Virginia Museum of Natural History) and published by The McDonald & Woodward Publishing Company in 1991. This shorter version is designed to make information on endangered and threatened species more readily available to all citizens of the Commonwealth by using non-technical language whenever possible, by using English units of measurement instead of metric, by eliminating many of the scientific details of species identification, by focusing only on Virginia distribution areas and habitats, and by significantly reducing the purchase price of the book.

Unlike its parent volume, *A Guide to Endangered and Threatened Species in Virginia* contains information only on those species in the state that appear on or have been formally recommended for inclusion on official federal or state lists as endangered or threatened. In many ways, thus, it only presents the tip of the iceberg; many more species are considered rare and vulnerable to extinction.

Since publication of *Virginia's Endangered Species,* field studies have led to status changes and even the discovery of species previously unknown in Virginia (for example, Michaux's sumac). Running glade clover was only described as a species in 1992. Thus there do appear several species which were not included in the original volume (in addition to the plants just mentioned, three species of freshwater mussel and the roseate tern) and several revised or expanded species accounts for newly listed species (for example, the Lee County cave isopod). It was also decided to give listed marine mammal species individual species accounts in this new book instead of the collective treatment afforded in the parent volume. Still, because of its more in-depth discussion of species, their natural history, and their distributions and of the threats to their continued existence in Virginia, *Virginia's Endangered Species* should remain the primary scientific reference for rare and vulnerable species in the state.

All abridged species accounts and chapter introductions were sent to the original authors for their review. Many were returned with new or revised information based on a few more years of research and analysis. While we hope that the book contains all the latest information on all species, the reader should recognize that new information is always becoming available and this book by no means represents the last word on these species or their status in Virginia. Indeed, it is hoped that publication of this book will contribute to the expansion of our knowledge of and caring for our most vulnerable co-inhabitants of the Commonwealth by allowing and encouraging more people to participate in their conservation. Success in conservation will be marked when a new volume is needed — not because there are more species to add, but because some may be deleted, their status improved, their future more secure, and their names removed from state and federal lists.

I would like to express my thanks to all the authors and chapter chairs who took the time to read and comment upon my abridgments of their previous writings. I also want to thank those who produced new accounts especially for this book. Most of all, I would like to express my appreciation to Karen Terwilliger of the Nongame and Endangered Wildlife Program of the Virginia Department of Game and Inland Fisheries, who once again coordinated efforts and exemplified the cooperative spirit needed if we are all to make a difference and conserve our threatened and endangered species.

Susan Woodward, editor
Department of Geography
Radford University

Introduction

Virginia's flora and fauna are among the most diverse to be found anywhere in the temperate latitudes of the earth. Yet as Virginia is transformed into an almost purely cultural landscape, the rich natural history of the Commonwealth is increasingly threatened; many native species of plants and animals stand to be greatly reduced in numbers or even eliminated.

Citizens who directly utilize biological resources of economic, recreational, or aesthetic value such as productive soils, native vegetation, wild game, or estuarine fin- and shellfish have always had a vested interest in the supply and availability of those resources. Formal efforts to provide legal protection to natural resources in Virginia date from as early as 1680, when legislation was enacted to prevent the wasteful harvest of fish in the Rappahannock River. Environmental protection now takes the form of legal mandates to state, local, and federal governmental agencies to regulate the use and effect the management and study of the biotic resources of the Commonwealth.

State Entities Responsible for the Conservation and Protection of Virginia's Endangered and Threatened Species

Two state agencies, the Department of Game and Inland Fisheries and the Department of Agriculture and Consumer Services, have authority to legally protect endangered and threatened species and are responsible for their conservation in Virginia.

Virginia Department of Game and Inland Fisheries

The mission of the Virginia Department of Game and Inland Fisheries is to manage Virginia's wildlife and inland fish to maintain optimum populations of all species to serve the needs of the Commonwealth; to provide opportunity for all to enjoy wildlife, inland fish, boating and related outdoor recreation; and to promote safety for persons and property in connection with boating, hunting, and fishing. Its vision is to manage a diversity of fish and wildlife species; provide environmental protection; enhance hunting, fishing, and boating programs; and provide wildlife-related recreational opportunities.

Since its establishment in 1916, the Virginia Department of Game and Inland Fisheries has been responsible for "conserving, protecting, replenishing, propagating, and preserving the wildlife of the lands and inland waters of the Commonwealth" (§29.1–103). Conservation of both rare and abundant wildlife is accomplished through a combination of survey and inventory, research, and management practices combined with education, enforcement, and regulation processes. Land acquisition, leasing, cooperative agreements, and a variety of other land conservation mechanisms are employed to secure wildlife habitat and to provide access for recreational opportunities focused on wildlife.

State law §29.1–521, enforced by the Virginia Department of Game and Inland Fisheries, provides blanket protection to all wild animals in the Commonwealth by prohibiting their possession and capture, except as permitted. General permitting allows specialized and restricted use such as collecting specimens for scientific, museum, or educational purposes. In 1972, Virginia's Endangered Species Act was passed to provide specific protection to those species of wildlife most vulnerable to extirpation or extinction. This legislation prohibits the taking, transport, and sale of endangered and threatened species, except as permitted, and empowers the Board of the Virginia Department of Game and Inland Fisheries to adopt the federal list of endangered and threatened species and to establish and implement regulations to further protect these species. The definition of "take" in Virginia's regulation is identical to that now in the federal Endangered Species Act, where the term encompasses not only capture and killing, but also harm, including harm done through modification of habitat and through harassment.

In 1976, under Section 6 of the federal Endangered Species Act, the Virginia Department of Game and Inland Fisheries entered into a cooperative agreement with the US Fish and Wildlife Service, which gave the Department the lead role and authority for conserving federally endangered and threatened animals in the Commonwealth. The Department adopted the list of federal endangered and threatened animals as well as the federal definitions and created Virginia's state endangered species list in 1987.

The Department of Game and Inland Fisheries established a dynamic system for determining the status of rare species and their need for protection at the state level prior to their listing in 1987. An objective, quantitative ranking system to be used in conjunction with recommendations from taxonomic advisory committees was created to provide the rigorous procedures needed to maintain accurate and current status information on the rarest and most vulnerable animals. The department solicits input from experts and reviews and updates its data annually. Recommendations for listing are presented to the Board for inclusion on the list of state endangered and threatened species when sufficient documentation and staff and external expert testimony support the listing. Notice of Board meetings to entertain listing recommendations as well as resulting Board listing actions are published in the *Virginia Register* in accordance with the Administrative Process Act.

Endangered and threatened species conservation is implemented through six divisions of the Department. The Wildlife and Fisheries divisions develop and implement conservation and recovery plans; perform survey, research, and management; and provide technical expertise within the agency and to public and private entities and landowners. The Planning, Policy, and Environmental Services Division maintains a comprehensive computerized wildlife database (the Fish and Wildlife Information System), manages and disseminates this information, and incorporates it into the environmental review process and into agency planning and policy development. The Public Relations and Education Division provides information on agency programs through a wide variety of media and has developed environmental education programs for both adults and children. Project Wild and Aquatic Wild K–12 curricula, the two most popular programs, have reached most of the state's schools and educators. The Lands and Engineering Division utilizes numerous land protection tools to acquire or otherwise protect habitat for wildlife and for wildlife-related recreation. Lastly, the Law Enforcement Division, through its county game wardens and regional staff, provides the important link between wildlife education and the enforcement of the many laws and regulations of the Department which protect wildlife species.

Funding for the Department's programs has come primarily from sportsmen through license fees and taxes on hunting and fishing equipment. For more than 75 years the federal aid in wildlife and fish restoration acts, generally known as the Pittman-Robertson and Dingell-Johnson acts, respectively, have provided for the return of federal taxes on sporting equipment to state wildlife agencies. In 1981, when the Nongame Bill was passed by the Virginia Assembly, the Department which already had responsibility for conserving and protecting nongame and endangered species began to receive monies from the Nongame Fund, which derives its revenues from a state income tax checkoff as well as direct contributions. Federal Endangered Species Section 6 funds are also utilized when available.

Virginia Department of Agriculture and Consumer Services

In 1979, the Endangered Plant and Insect Species Act, Chapter 39 §3.1–1020 through 1030 of the Code of Virginia, as amended, mandated that the Virginia Department of Agriculture and Consumer Services conserve, protect, and manage endangered and threatened species of plants and insects. The impetus for and original intent of the legislation was to provide Virginia with federally required legislation to control the export of American ginseng. When the Act was passed, it listed ginseng as threatened and Virginia round-leaf birch as endangered.

The Endangered Plant and Insect Species Act empowers the Board of Agriculture and Consumer Services to prescribe and adopt rules and regulations to list species of plants and insects as threatened or endangered. Once a species is listed as threatened or endangered, it is unlawful for any person other than the landowner to dig, take, cut, process, or otherwise collect, remove, transport, process, sell, offer for sale, or give away individual plants except by special permit issued by the Commissioner of Agriculture and Consumer Services. Permits for possessing individuals of a listed species may be issued for scientific or educational purposes or for propagation in order to insure the survival of the species. In addition, permits to remove individuals or populations of listed species may be issued to alleviate damage to private property, to alleviate adverse impact on progressive development, or to protect human health.

The Endangered Plant and Insect Species Act also regulates the sale and movement of those endangered and threatened species under its purview. The taking of a threatened species is permitted if the Board of Agriculture and Consumer Services determines that the abundance of the species in Virginia justifies a controlled harvest and would not be in violation of federal laws and regulations. A licensing and record-keeping system has been implemented to control the harvest of threatened species.

The process for listing plant and insect species begins with experts' surveys of rare species. Once the status of a species is known, a written recommendation is made by the Director of the Department of Conservation and Recreation or other reputable authorities to the Commissioner of Agriculture and Consumer Services for listing under the Endangered Plant and Insect Species Act. The Commissioner then determines those species in most need of protection and presents them to the Board of Agriculture and Consumer Services as *candidate* species, requesting that the Board proceed with the Administrative Process Act and list the species under the Endangered Plant and Insect Species Act. The intent to develop regulations is published in the *Virginia Register*, and a public hearing is held before the Board votes on the proposed listing of the species. A majority vote is required to list a species as threatened or endangered. Once a species is listed, the Commissioner may begin the development and implementation of a management program for the protection and recovery of the species. Funding for the implementation of a recovery plan is provided by the Department and the US Fish and Wildlife Service.

Two other natural resource agencies, the Department of Conservation and Recreation and the Virginia Museum of Natural History, also contribute significantly to the conservation of endangered and threatened species and their habitats in Virginia.

Virginia Department of Conservation and Recreation

The mission of the Virginia Department of Conservation and Recreation is to conserve, protect, enrich, and advocate the wise use of Virginia's natural, recreational, and scenic resources in order to maintain and improve the quality of life for present and future generations. The Department supports a variety of environmental programs organized within five divisions (Administration, Natural Heritage, Planning and Recreation Resources, Soil and Water Conservation, and State Parks) and four policy boards (The Board of Conservation and Recreation, The Virginia Cave Board, The Board on Conservation and Development of Public Beaches, and the Virginia Soil and Water Conservation Board). It is one of the major land acquisition and management agencies in state government. The majority of its land is managed for multiple use and kept in a relatively undeveloped and natural condition. High priority is given to the acquisition and protection of open space and biologically significant habitat. Such properties are living outdoor classrooms in which present and future generations of Virginians can study and enjoy the Commonwealth's irreplaceable biological heritage.

The Natural Heritage Division, created in 1988, represents the first comprehensive attempt to identify the most significant natural areas in the Commonwealth through an intensive statewide inventory of plants, animals, natural communities, and other features that are exemplary, rare, or endangered on a global or statewide basis. The Virginia program is part of an international network of Natural Heritage programs spanning all 50 states as well as parts of Canada and Latin America; a consistent methodology allows information to be readily shared and compared to establish conservation priorities across state and national boundaries.

Four major programs compose the Natural Heritage Division: Natural Heritage Inventory, Data Management, Natural Areas Protection, and Ecological Management. The inventory has identified over 900 conservation sites consisting of one or more rare species and exemplary natural communities. The Biological and Conservation Data Management System represents a permanent atlas and database on the existence, characteristics, abundance, status, and distribution of Virginia's biodiversity; it tracks site-specific data on more than 8,000 locations for Virginia's rare species, communities, and ecosystems.

The Natural Areas Protection program began with the Natural Area Acquisition Fund, a cooperative effort between the Department of Conservation and Recreation and the Virginia Chapter of The Nature Conservancy. This alliance set in motion the acquisition of biologically significant areas identified by the then Virginia Natural Heritage Program and the creation of a Natural Area Preserves System, the first state land management program specifically designed to conserve and manage Virginia's rare, threatened, and endangered biological resources. Dedication of properties as Natural Area Preserves, which can be accomplished through the voluntary act of a landowner, provides strong statutory protection against conversion to alternate uses. Acquisition of Natural Area Preserves is often pursued with funds from the Natural Area Preservation Fund and the Open Space Recreation and Conservation Fund, the latter consisting of revenues derived from voluntary contributions through a state income tax checkoff. The importance of protecting these sites was endorsed by the voters in their overwhelming support of the 1992 Park and Natural Areas Bond, which provided $12.1 million to purchase at least ten new Natural Area Preserves.

Stewardship, or land management, is the responsibility of the Ecological Management Program. Proper management ensures the continued existence and enhancement of significant natural areas and natural heritage resources through careful monitoring, scientifically based ecological studies, and prescribed management actions. The Division's ecologists and land managers offer their technical expertise in natural area management not only to the Department, but also to other state agencies and to local and federal agencies and private landowners and land managers.

Virginia Museum of Natural History

The Virginia Museum of Natural History, established in April 1988 as a state agency under the Secretary of Natural Resources, was designated an "instrumentality for the purpose of preserving and protecting Virginia's natural history." Among its several specific missions are the expansion and dissemination of existing knowledge of Virginia's biodiversity through educational activities and the generation of new information through original research conducted in eight scientific departments.

The museum conducts biotic surveys and participates in such surveys conducted jointly with other state agencies, provides permanent storage for voucher and other specimens taken during these surveys, develops reference collections in most animal groups, provides information on the genetic viability of isolated populations, and emphasizes environmental and conservation awareness through its exhibit and educational outreach programs. The museum's exhibit department is licensed at both federal and state levels to receive, store, and prepare specimens of listed species.

Federal Legislation Aiding the Conservation and Protection of Virginia's Plants and Animals

Many federal laws aid the conservation and protection of plants and animals occurring in Virginia. Among those not already mentioned are the Migratory Bird Treaty Act of 1918, which protects birds that migrate between the US and Canada; the Migratory Bird Hunting and Conservation Stamp Act of 1934, which provides revenues from a federal "Duck Stamp" for the acquisition and management of migratory bird refuges; the Bald Eagle Act of 1940, which protects both the bald eagle and the golden eagle; the Fish and Wildlife Act of 1956, which established the US Fish and Wildlife Service and its program of research, extension, and information services; and the Marine Mammal Protection Act of 1972, which established a moratorium on the taking and importation of all endangered marine mammals and products made from them.

The first federal listing of rare and endangered wildlife appeared in 1966 as the "Red Book" or Federal List of Rare and Endangered Fish and Wildlife of the United States. It was produced by the then Bureau of Sport Fisheries and Wildlife. The federal Endangered Species Act was enacted in 1973 and sought "to provide a means whereby the ecosystems upon which endangered species depend may be conserved, to provide a program for the conservation of such endangered species and threatened species, and to take such steps as may be appropriate to achieve the purposes of the treaties and conventions" in which the United States has pledged its support for the conservation of wild flora and fauna worldwide.

The law established two categories of endangerment: (a) *Endangered Species* — those species in danger of extinction throughout all or a significant portion of their range; and (b) *Threatened Species* — those which are likely to become endangered within the foreseeable future throughout all or a significant portion of their range. This law also emphasizes the need to preserve critical habitats upon which endangered species depend for their continued existence. Individual states were encouraged to establish guidelines to complement the goals outlined in the 1973 Act. Virginia responded to that by developing a cooperative Section 6 agreement by the Virginia Department of Game and Inland Fisheries for all animals except insects and the Virginia Department of Agriculture and Consumer Services for plants and insects. These agreements provide not only the necessary authority but also the framework for the cooperative and effective implementation of the federal Endangered Species Act at the state level by these two state regulatory agencies.

Some federal legislation not directly aimed at species conservation provides important environmental protection that benefits threatened and endangered species. Such laws as the Wilderness Act (1964), the Wild and Scenic Rivers Act (1968), National Environmental Protection Act (NEPA) (1969), the Water Bank Act (1970), and the Clean Water Act (1972) contribute to the conservation of Virginia's and the nation's natural history legacy.

How You Can Help

Sound scientific inquiries and methods provide the data that make possible the credible expression of the popular will by governmental entities, but the formation of public, private, and scientific partnerships has enhanced opportunities for achieving truly successful protection and restoration of Virginia's natural resources. Both state and federal agencies responsible for endangered and threatened species protection recognize the tremendous value of the involvement of all people and strongly support the efforts of the citizenry, special interest groups, private landowners, and the private sector towards the development of comprehensive conservation programs.

There are many ways the average citizen can help conserve and protect wild species. Below are just ten things you can do to help.

- Learn and teach others: recognize plant and animal species and understand basic ecological principles.
- Encourage proper land use practices in your community and practice them yourself.
- Support public and private conservation initiatives:
  - Support wildlife conservation issues.
  - Be a member of local, state, national, or international conservation organizations.

- Use the income tax checkoffs for the Nongame and Endangered Wildlife and the Open Space Conservation and Recreation funds.
- Purchase a Duck Stamp and hunting and fishing licenses each year even if you don't hunt or fish, since many of the collected revenues go to species and habitat conservation.
- Obey wildlife and, in particular, endangered species laws and regulations and report violations.
- Use wildlife refuges, parks, and nature preserves and all habitats in ways that minimize the impacts of your presence: stay on trails, take only pictures, and leave only footprints.
- Conserve water and energy.
- Do not litter or pollute, but do help clean up.
- Dispose of hazardous household and commercial wastes properly.
- Recycle whenever possible.
- Reduce and mitigate the use of fertilizers and pesticides on gardens and lawns as well as farmland, use these chemicals properly, and explore alternative organic farming and pest control techniques.

Based on "Introduction" by Karen Terwilliger, pp. 3-22 in *Virginia's Endangered Species*, coordinated by Karen Terwilliger, 1991 (Blacksburg, VA: The McDonald & Woodward Publishing Company).

The Nature of Virginia

Virginia lies on the east coast of the United States at the poleward edge of the subtropics. In latitude, the state reaches from 36° 30' N along it southern border to about 39° 30' N at its northernmost point. Virginia's location and its varied physical environments have had significant influence on the establishment of species-rich plant and animal communities.

Physiographic Provinces

With its long east-west axis, Virginia intersects five of North America's more than 20 physiographic provinces — landform regions that reflect unique histories of rock formation, deformation (or lack thereof), and subsequent erosion. These five provinces provide a multitude of different microenvironments for life to exploit and have a major influence on the high diversity of life found in the state.

Virginia's physiographic provinces (Figure 1) represent parts of two major landform units of the North American continent, the Atlantic Coastal Plain and the Appalachian Highlands. The Atlantic Coastal Plain is geologically much younger than the Appalachian Highlands unit and consists of the Coastal Plain Province and the Continental Shelf, the submerged eastward extension of the continent. The Appalachian Highlands comprises four provinces: the Piedmont Plateau, the Blue Ridge, the Ridge and Valley, and the Appalachian Plateaus.

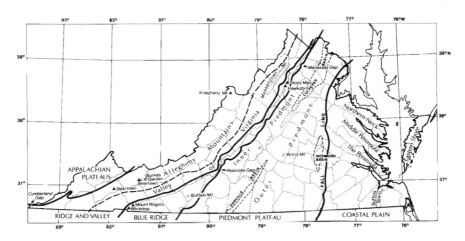

Figure 1. The physiographic provinces of Virginia, with locations of major subregions and landform features.

9

Coastal Plain Province

The Coastal Plain is built of poorly consolidated sediments eroded from the Appalachian Highlands; underlying layers of sedimentary rock tilt toward the sea. Fluctuating sea levels related to tectonic and climatic changes have left a series of wave cut terraces as the major topographic features of the plain.

The Coastal Plain is a surface of low relief which slopes gently to sea level from a maximum elevation of about 200 feet above sea level (asl) at the Fall Line. The province is by no means homogenous. It is dissected into four peninsulas, commonly designated the Northern Neck, the Middle Peninsula, The Peninsula, and the Eastern Shore. Minimum relief occurs on the Eastern Shore. On the mainland there is a pronounced change of surface character from north to south. The relatively narrow Northern Neck has been rather deeply dissected by short, high-gradient streams into a terrain justly described as hilly and well-drained. Southward the surface gradually becomes flatter until, south of the James River, it is virtually featureless except for the several wave cut terraces and the ravines developed upon them. Along the shores of the Chesapeake Bay and the Atlantic Ocean, both exposed and protected beaches and both active and stabilized dunes add variety to the landscape. Such diversity, although subtle, accounts for the development of a variety of plant and animal communities on the Coastal Plain.

The Continental Shelf, a part of the Coastal Plain, extends 50 to 75 miles eastward from the shoreline to a depth of 600 feet below sea level at the edge of the North American continent. The shallow waters above it receive nutrients from the land and are home to rich wetland and open sea communities.

Piedmont Plateau Province

The Piedmont is a gently rolling upland bounded on the east by the Fall Line and on the west by the escarpment of the Blue Ridge. In the northern parts of the state the Piedmont is only about 50 miles wide, but it broadens in the south to a width of about 150 miles. It begins at about 200 feet asl at the Fall Line and rises to approximately 1,000 feet asl in the west.

In places, distinct peaks and discontinuous ridges stand hundreds of feet above the general surface of the Piedmont. They are monadnocks, remnants of a former higher surface and composed of rock more resistant than that of the surrounding lowlands. Most monadnocks lie in the western part of the province. An imaginary line along the eastern edge of the monadnock region separates the rugged Inner Piedmont from the more gently rolling Outer Piedmont to the east.

The usually low relief of the Piedmont belies a complex geological history wherein folding, faulting, metamorphism, and uplift occurred. Much of the bedrock consists of resistant slates, gneisses, and schists. In places, intrusive granite outcrops but has been eroded to low features known as granite flatrocks.

The Piedmont was affected by rifting when the Atlantic Ocean formed. Great blocks of the earth's crust were downfaulted, creating long rift valleys parallel to the edge of the continent. Magma intruded along the faults and sediments eroded from the Appalachian Highlands to the west have filled the valleys to create today's flat Triassic lowlands.

Blue Ridge Province

Geologically, the Blue Ridge Province represents a slice of the continent's basement rock which was faulted and lifted over younger rocks as the Appalachian Mountains formed. Ancient igneous rock as well as overlying metamorphic rock related to that of the Piedmont Plateau crop out in the province. Resistant granites, greenstones, and quartzites form a terrain of high relief. The young, fast-flowing streams which drain to the Atlantic have rapidly eroded the uplands in the northern part of the Blue Ridge so that only a narrow chain of peaks less than five miles across remains. These are the Blue Ridge Mountains. South of Roanoke Gap, the Blue Ridge is drained by the lower gradient streams of the New River system, and it broadens into a plateau up to 50 miles wide. The surface elevation of the Blue Ridge Province generally increases from approximately 2,000 feet asl in the north to an average of 3,000 to 3,500 feet asl in the south. The highest peak in the Blue Ridge Mountains is Hawksbill Mountain at 4,050 feet asl, while the southern Blue Ridge Plateau of Virginia is crowned by the two peaks of Balsam Mountain, the two highest points in the state: Mount Rogers at 5,729 feet asl and Whitetop at 5,520 feet asl. Both peaks are composed of volcanic rock and are lithologically distinct from other rock materials of the Blue Ridge.

Ridge and Valley Province

Most of western Virginia lies in the Ridge and Valley Province. Long parallel ridges and intervening valleys characterize the province, which is underlain by folded sedimentary rocks of Paleozoic age. The long axes of folds control the alignment of the ridges and valleys, but the differential erosion of rocks of varying resistance to a large degree has determined contemporary landform patterns. Where resistant sandstones are at the surface, they form long, narrow mountain ridges or broader uplands. Less resistant carbonate rocks (limestones and dolomites) and shales, on the other hand, have eroded more rapidly and tend to form lowlands and long valleys paralleling the sandstone ridges. Carbonate rocks dissolve to form a type of topography known as karst. Sinkholes pit the surface of most karst areas, and below ground caves and caverns develop along joints and bedding planes.

The eastern half of the Ridge and Valley Province is part of the Great Valley, a feature that extends along the western flank of the Blue Ridge from the Hudson River in New York to northern Alabama. In Virginia, this broad valley is known as the Valley of Virginia and is subdivided into a number of segments named according to the major stream of the region. In the north, the Shenandoah Valley slopes from about 600 feet asl at the Virginia-West Virginia border to 1,400 feet asl at the James River basin divide. Towering 1,000 to 1,700 feet above the valley floor and dividing it and the forks of the Shenandoah River is 50-mile long Massanutten Mountain.

To the south, the Roanoke Valley has an elevation of about 1,000 feet asl, but southwest of Roanoke the Valley of Virginia becomes both higher and narrower. The New River Valley is 800 to 1,000 feet higher than the adjacent Roanoke Valley; and the Holston Valley drops from 2,500 feet asl at Rural Retreat to about 1,600 feet asl at Bristol.

A belt of long, narrow, parallel mountains and valleys, the Allegheny Mountain section of the Ridge and Valley Province, lies west of the Valley of Virginia. The highest peak in Virginia's Alleghenies, Beartown Mountain (Tazewell County), reaches about 4,700 feet asl.

Appalachian Plateaus Province

The far southwestern counties of Virginia lie wholly or partly within the Cumberland Plateau section of the Appalachian Plateaus Province. The province differs from the Ridge and Valley primarily by the fact that the sedimentary rocks were never deformed by mountain-building activity and remain essentially horizontal. The modern landscape evolved as small streams in tree-like patterns incised themselves into the former plateau surface and created a mountainous topography. The Cumberland Plateau today is an intricately dissected land where level surfaces are rare. Summits of hills are 2,000 to 2,500 feet asl; the valleys and hollows are 500 to 1,000 feet lower.

Drainage Systems

The eastern continental divide passes through Virginia dividing it into two large drainage basins. Most of the land west of Roanoke, roughly one fourth of the state's surface area, drains to the Gulf of Mexico via the Big Sandy, Tennessee, and New-Kanawha river systems. Streams of the remaining three-fourths of the state flow generally eastward into the Atlantic Ocean (Figure 2).

In the Appalachian Plateaus Province, Russell Fork (which leaves the state at The Breaks in Dickenson County) and the Levisa Fork (which enters Kentucky northwest of Grundy, Buchanan County) are parts of the Big Sandy drainage system. Both streams are deeply entrenched in steep-sided canyons.

The Powell, Clinch, and Holston rivers, major tributaries of the Tennessee River, drain the Valley and Ridge Province of southwestern Virginia. The Tennessee River system and the neighboring Cumberland River system in Kentucky are home to an exceptionally diverse fish and mollusk fauna known collectively as the Cumberlandian fauna. Much of the fish and mollusk diversity found in Virginia's streams is due to direct links — past or present — with these two river systems.

The New River begins in the Blue Ridge of North Carolina, flows north and northwest across the Blue Ridge and Ridge and Valley provinces of Virginia to enter the Appalachian Plateaus Province in West Virginia. There, near Charles-

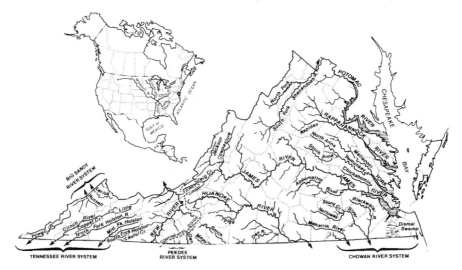

Figure 2. The surface drainage systems of Virginia.

ton, it is joined by the Gauley River and becomes the Kanawha, which drains via the Ohio and Mississippi rivers to the Gulf of Mexico. The New River is believed to pre-date the mountain-building episode that formed the Alleghenies and is considered the oldest river on the North American continent. The New River basin's eastern edge is the eastern continental divide. The higher gradient streams of the Atlantic Slope have continually cut westward, capturing tributaries of the New River and reducing its drainage area. Stream piracy has provided passage for aquatic species of the Ohio River basin into the Atlantic drainages.

Seven river systems make up the Atlantic Slope drainage of Virginia. The southernmost of these, the PeeDee, is of such minor influence in this state that it will not be discussed here. The remaining six may be grouped into those that empty into Albemarle Sound in North Carolina (the Roanoke and Chowan river systems) and those that enter the Chesapeake Bay (James, York, Rappahannock, and Potomac river systems).

A large portion of the southern Piedmont is drained by the Roanoke River and its southern branch, the Dan River. The upper Roanoke River has cut through the Blue Ridge and some 30 miles of the stream occur in the Ridge and Valley Province in Roanoke and Montgomery counties. Topography and distribution of fishes imply that the Roanoke and Dan rivers have captured tributaries of the New River west of the Blue Ridge. Many of the Roanoke's diverse and endemic fishes are restricted to the upland habitats of its western headwater streams.

The Chowan River system is represented in Virginia by the Meherrin, Nottoway, and Blackwater rivers. The Meherrin in its upper third is deeply entrenched and has a substantial gradient. Downstream it becomes deeper and straighter until, below the Fall Line at Emporia, it forms a highly convoluted pattern of meanders and oxbows. Throughout much of their lower courses both

13

the Nottoway and Blackwater are slow, entrenched streams with deep channels. The Blackwater flows through cypress swamps, while the Nottoway is embellished with swift, sandy-bedded reaches of great beauty. Before merging to form the Chowan River in North Carolina, both streams become estuarine.

The James River has its headwaters in the central Allegheny Mountains of Virginia. It meanders through the Valley and Ridge to the town of Buchanan. There it abruptly turns northward along the western base of the Blue Ridge, which it subsequently crosses through the spectacular gorge at Balcony Falls. Once on the Piedmont, the James River acquires only a few modest tributaries. Immediately below the Fall Line at Richmond, the James River turns southward and becomes estuarine. The Appomattox and Chickahominy rivers are the main tributaries entering the James on the Coastal Plain.

The York River drainage basin lies primarily on the Outer Piedmont. The York River itself, formed by the convergence of the Pamunkey and Mattaponi rivers, is entirely estuarine.

Most of the headwater streams of the Rappahannock River begin on the Blue Ridge. The Rapidan River is its major tributary. Below the Fall Line, there are virtually no tributaries and the basin narrows appreciably. The Rappahannock is tidal for more than half its length.

With the exception of the headwaters of the South Branch, which arise in northern Highland County, the Potomac River proper lies wholly outside of Virginia. In 1632, the King of England granted the Potomac River to Lord Baltimore and the south bank of the Potomac became Virginia's northern boundary. The most significant of the Potomac's south bank tributaries is the Shenandoah River, which empties into the Potomac at Harpers Ferry, West Virginia.

The Potomac, Rappahannock, York, and James rivers were linked as tributaries of the Susquehanna River during periods of low sea level in the Pleistocene. Interbasin migration of lowland fish species was then possible, so today the fish fauna of the James River is much more similar to that of the Potomac River than to that of the adjacent Roanoke River, which was never connected to the Susquehanna system.

Climate

Virginia's climate is classified as humid subtropical, a climate type characterized by warm to hot summers and mild winters. Topography and elevation have major impacts on both temperature patterns and precipitation patterns across the state. In addition, the moderating influence of the sea on temperature is experienced near the coast, especially in autumn and winter. The longest growing season (259 days) is recorded at Cape Henry, City of Virginia Beach, while the shortest growing season (135 days) recorded in the state occurs in Burkes Garden high in the Allegheny Mountains.

Virginia has an average annual precipitation of 42 inches, but there is considerable regional variation. The Shenandoah Valley, for example, receives an average of 34 inches a year — equivalent to that of the eastern prairies; while the far southwestern part of the state averages more than 50 inches a year. Usually less than 10 inches of snow falls each year on the Coastal Plain and less than 20 inches west of the Blue Ridge. Except on the higher mountains, snow rarely persists more than a few days anywhere in the state.

Natural Vegetation

A temperate broadleaf deciduous forest covers most of Virginia, reflecting its humid subtropical climate. Various types of this forest may be found from northern New England to southern Florida, from the Mississippi River to the Atlantic Ocean. In Virginia, the forest is generally characterized by four layers, each rich in species. The canopy, dominated by mast-producing oaks and hickories, is 60 to 100 feet above the forest floor. Below it lies an understory of smaller trees such as dogwood and redbud; a shrub layer frequently dominated by heaths such as rhododendron, azalea, and mountain laurel; and an herb layer of diverse perennial forbs, mosses, lichens, and clubmosses. Woody vines are conspicuous in moister habitats; most common are wild grape, Virginia creeper, and poison ivy.

Virginia's broadleaf deciduous forest may be differentiated into four basic types: mixed mesophytic, oak-chestnut, oak-pine, and southeastern evergreen forests. Most diverse is the mixed mesophytic forest found in the Appalachian Plateaus Province. Although the member tree species tend to segregate into a mosaic of stands, more than 20 species are considered to share dominance in this forest type. The species include American beech, sugar maple, eastern hemlock, red oak, white basswood, tulip tree, yellow buckeye, and various hickories, ashes, and magnolias. Outliers of this forest type, known as cove forests, occur in cool, damp, deep valleys in the Allegheny Mountains and Blue Ridge.

The most widespread forest type in Virginia is the oak-chestnut forest, which covers most of the Ridge and Valley, Blue Ridge, and northern Piedmont Plateau provinces. Three oak species are most frequent: white oak, chestnut oak, and red oak. Hickories are also important components. The American chestnut, once the largest and perhaps most important tree in the forest, has been extinct as a canopy species since the 1930s.

In the Blue Ridge and the Ridge and Valley, the oak-chestnut forest varies with increasing elevation. A subtle altitudinal zonation can be discerned between 4,000 and 4,500 feet asl as oaks and hickories yield dominance to American beech, sugar maple, and yellow birch, northern hardwoods more typical in the northeastern US. Above 4,500 feet, most notably on Beartown Mountain (Russell County), Whitetop, and Mount Rogers, the broadleaf forest gives way entirely to a needleleaf evergreen forest. In this uppermost (boreal or Canadian) zone, red spruce usually dominates, but on Mount Rogers a southern Appalachian endemic, Fraser fir, covers the summit.

On the southern Piedmont and the peninsulas of the Coastal Plain, pines become more abundant and black oak replaces red oak as the principal co-dominant with white oak in the region's oak-pine forest. Virginia pine and shortleaf pine are common. On the Coastal Plain and the eastern edge of the Piedmont, these two short-needled pines are joined by the long-needled loblolly pine. Pines occur primarily as members of early successional communities on abandoned farmland, but on dry sites and on soils with low nutrient content — such as those exhausted from poor agricultural practices, pines may persist.

The southeastern evergreen forest occurs on the Coastal Plain south of the James and is the northernmost extension of a vegetation type in which long-needled pines dominate. This forest stretches southward and westward from Virginia to eastern Texas. Longleaf pine is characteristic but generally confined to sandy uplands, where it is maintained by low nutrient, well drained sandy soils and periodic fire. Where drainage is poor, loblolly pine and pond pine join longleaf pine in a savanna with an herb layer of grasses, sedges, and flowering forbs. On heavier, alluvial soils along rivers and in Great Dismal Swamp, a swamp forest characterized by bald cypress and dominated by tupelo, red maple, and black gum occurs. At maritime sites such as Seashore State Park, cypress may be accompanied by live oaks heavily festooned with Spanish moss.

Restricted and Vulnerable Communities

Within Virginia, there are several unique, localized communities more strongly influenced by local microclimates and substrate than by the regional climate. Most often it is in such habitats that Virginia's rarest and most vulnerable plant and animal species are found.

High Elevation Communities

The cooler temperatures and higher humidities of high elevations permit mountain summits to serve as refuges for species which had much broader distributions in Virginia during the Pleistocene, species with boreal and even arctic affinities. Two high elevation habitats, boreal forests and bogs, are particularly significant in harboring northern species.

The boreal forest is exemplified by the red spruce-Fraser fir forest atop Mount Rogers. Fraser fir occurs naturally in no other boreal stands in Virginia. Elsewhere at high elevations red spruce is the dominant. In the spruce forests there is often a dense understory of rhododendrons. Within some communities may be found the endangered snowshoe hare and northern flying squirrel.

Bogs are cool, acidic, and extremely small wetlands that occur in areas of impeded drainage at high elevations. They have microenvironments similar to the true bog or muskeg found farther north in glaciated parts of North America. Virginia's bogs occur primarily in the New River drainage section of the Blue Ridge and on Massanutten, Clinch, and Salt Pond mountains in the Ridge and Valley. Sphagnum moss is characteristic of these wetlands. Rushes, sedges, grasses, and various forbs and subshrubs make up the rest of the community. Virginia's bogs, unlike true muskeg, contain not only species of northern affinities but also species of southern affinity and some disjunct coastal species. The endangered bog turtle may be found in some bogs of the southern Blue Ridge.

Freshwater Communities

Virginia has only two natural lakes, Mountain Lake in Giles County and Lake Drummond in the Great Dismal Swamp. Neither is inhabited by listed species. However, Virginia's streams, pools, ponds, and swamps provide critical habitat for a host of rare species, among them plants, mollusks, arthropods, fishes, amphibians, and even a few mammals.

Any drainage system includes diverse microenvironments ranging from fast, cool upland headwaters to slower, deeper lowland stretches near the mouth. Differing substrates over which waters flow further contribute to the diversity of microhabitats.

Freshwater swamps are habitats where woody vegetation grows in standing water for all or much of the year. Swamps may be found in all the physiographic provinces, and each one is a critical habitat. The largest swamp in Virginia is the Great Dismal Swamp, which spreads over some 750 square miles atop impermeable clays on the Coastal Plain of southeastern Virginia and northeastern North Carolina. It is critical habitat for several vulnerable mammals, including the threatened Dismal Swamp subspecies of southeastern shrew.

Another significant but restricted freshwater habitat is the evergreen shrub bog or pocosin. Pocosins occur on sandy or peaty soils on flat, poorly drained ridges. Pond pine dominates an open-canopy tree layer that rises above a dense shrub layer of swamp magnolia, loblolly bay, holly, and red bay. Many of the shrubs are flammable and fire helps maintain the pocosin. The endangered canebrake rattlesnake occurs in this habitat.

A unique ephemeral freshwater habitat occurs in sinkhole ponds in karstic areas of several counties in the upper Shenandoah Valley. Sinkhole ponds lie above the limestones which have dissolved at depth to form pits that the ponds fill. The pond margins are actually acidic environments supporting wetlands conducive to the survival of both northern and coastal plant species. Among listed species, northeastern bulrush and Virginia sneezeweed grow in this habitat. Swamp pink occurs near some of them.

Substrate-Controlled Communities

In the Ridge and Valley Province, the nature of the substrate can have profound influence on the plant and animal communities developed on or within it. Particularly significant are carbonate bedrock and south-facing shale outcrops.

The dissolution of carbonate rocks has created more than 2,300 natural caves in Virginia. Many are home to a specialized and extremely localized fauna. Aquatic invertebrates, especially amphipods and isopods, have adapted to the cave environments, and several species of bats utilize caves for hibernacula and nurseries.

17

Steeply sloping (20° or more) shale outcrops, particularly when undercut by a stream, produce a very unstable surface of thin flakes of shale. Gravity continually pulls the flakes downslope and prevents the establishment of large trees. South-facing outcrops receive the most intense rays of the summer sun, adding high temperatures and high evaporation rates as limitations to plant growth. Such hot, dry exposures are known as shale barrens and upon them a sparse, shrubby pine-oak community is characteristic. Shale barrens are restricted to the Ridge and Valley Province and extend only from Virginia and West Virginia through Maryland to southeastern Pennsylvania. The few plant species which survive the harsh conditions are frequently endemic to the shale barrens or represent widely disjunct populations of primarily midwestern or southwestern species. The endangered Millboro leatherflower is an example of a shale barren endemic.

Coastal Communities

Virginia has two coastal regions, the Atlantic Coast proper and the Chesapeake Bay. Where the land meets the Atlantic head-on, the actions of waves and wind create dynamic landscapes that can change overnight with the onslaught of a winter storm or a late summer hurricane. The shore of the Bay, on the other hand, is shaped primarily by stream action from the great rivers which empty into it.

On the Atlantic Coast, barrier islands provide a broad zone of interface between land and open sea down the length of the Eastern Shore; but along the coast of Virginia Beach, the boundary is abrupt. Sandy beach and dune habitats exist in both locations. In Accomack and Northampton counties on the Eastern Shore, the dynamic but fragile areas of sand are nesting sites critical to the survival of several of Virginia's rarest birds, including piping plover, Wilson's plover, and gull-billed tern. The loggerhead turtle uses the barrier island beaches and those of Virginia Beach for nest sites. And the beaches of Mathews County on the Chesapeake shoreline are home to the state's four populations of the northeastern beach tiger beetle.

Virginia's coastal wetlands grade from saltmarshes secured behind the barrier islands to brackish marshes bordering Chesapeake Bay and edging the estuarine stretches of streams entering the Bay, to upstream freshwater marshes still influenced by the ebb and flow of the tides. Two communities are distinguished within the saltmarsh. High marsh is flooded only irregularly, during the highest tides of the year. Here grow saltmeadow hay, blackneedle rush, and salt grass. Low marsh is flooded daily at high tide and is dominated by saltmarsh cordgrass. Glassworts and sea lavenders also occur in the low marsh.

Considerable quantities of decaying vegetation accumulate in the saltmarsh and support a rich community of detritus-feeding invertebrates and their predators. The low marsh and its tidal streams also provide habitat for larval stages of insects and for mollusks, crustaceans, and fishes and thus are important habitat for a variety of terrestrial animals that come to the marsh to feed.

Tidal freshwater marshes occur along Coastal Plain rivers near the Fall Line; major freshwater marshes also fringe Back Bay and extend up North Landing River. Cattails, chairmaker's rush, Olney's three-square, and big cordgrass are significant members of the plant community. Tidal freshwater marshes in many places lie adjacent to the spawning grounds of estuarine and oceanic fishes such as shad, river herring, striped bass, and the Atlantic sturgeon.

At the edge of the land, below the low tide mark, in shallow brackish waters is a community of rooted grasses rich in marine invertebrates, the seagrass meadow. Eelgrass and associated plants trap nutrients and plankton circulating in the currents and provide food for diverse isopods, amphipods, snails, and sea slugs. The seagrass meadow also provides shelter for young fishes and for molting crabs. Eelgrass is consumed by the endangered Atlantic green and loggerhead sea turtles and by numerous migratory waterfowl which winter in the Chesapeake Bay area. The sea grass meadows are vital not only to Virginia's species but to others which breed or spend much of their adult lives far beyond Virginia's borders.

Human Alteration of Habitats

People first arrived in Virginia no later than 12,000 years ago, when an open spruce and pine forest covered most of the land area. Along with the climatic and vegetational changes associated with the retreat of the continental ice sheets to the north, Paleo-Indians were part of the great environmental revolution which resulted in the extinction or extirpation of Virginia's large Pleistocene mammals: mammoths and mastodons, musk oxen and caribou, ground sloths and horses.

With the demise of the largest game species, a warming climate, and re-establishment of the broadleaf forest, aboriginal populations apparently came to exploit a wider range of natural products from forest, stream and estuary as part of a generalized hunting and gathering economy that persisted throughout Virginia from roughly 9,000 to 3,000 years ago.

The widespread adoption of agriculture began about 3,000 years ago in Virginia. People cultivated maize, beans, and squashes. Agriculture was especially important to the tribes of Virginia's Coastal Plain, where by AD 1600 a relatively dense population lived in more than a hundred villages distributed along the major rivers. These people used fire to clear small plots of forests for croplands and apparently also regularly burned the forests to create meadows attractive to large game such as deer and elk and possibly bison. Fire was used to drive deer during the fall hunt. Repeated burning had opened the forest to the degree that the seventeenth century accounts by Europeans described woods through which one could easily ride a horse.

It is difficult to measure the true impact of Native Americans on Virginia's ecosystems. Peoples who came across the Atlantic — from Europe and Africa — created a new culture and a new landscape, obliterating the record of Indian environmental change. To the Europeans, Virginia was a vast wilderness to be tamed and civilized — "Europeanized" — and that meant converting forest to farmland.

In the process of taming the American landscape, they overharvested game and furbearers, exterminated predators, and cut down forests.

Widespread clearing of the forest marked the progression of European agriculture and people from the Coastal Plain onto the Piedmont during the first two centuries of settlement. Tobacco, which rapidly withdrew nutrients from the topsoil, was grown in a slash-and-burn system not unlike that of the Native Americans. While land was abundant and cheap, it was easy to move on when the soils were exhausted and clear new lands, leaving behind an untidy landscape of abandoned fields, frequently invaded by stunted pines.

Animal husbandry practices probably were responsible for modification of the forest. Cattle and hogs, left to forage for themselves, roamed the woods. Both livestock species must have altered the amount of cover and relative abundance of plants in the shrub and herb layers.

As the human population grew, lumbering increased. Wood was the primary construction material and fuel. Soap-making, glass-making, tanning, and wool-cleaning all required wood too. Fencing — to keep livestock out of cultivated plots — consumed considerable timber: the movable, self-supporting worm or Virginia fence was typically 6 to 10 rails high.

In the 1700s, attracted by rich soils and cheap land, Scotch-Irish and German-speaking peoples moved southward out of Pennsylvania into the Valley of Virginia and began clearing the forests west of the Blue Ridge for their farms. Early industrial development in the Ridge and Valley also depleted forests; iron furnaces demanded charcoal and tanneries demanded oak and chestnut bark. Only the most inaccessible areas of the Blue Ridge and Alleghenies escaped the axe.

It is safe to say that virtually all of the broadleaf deciduous forest in Virginia today is secondary growth, much of it less than 100 years old. Major periods of farm abandonment occurred in the northern Valley and northern Piedmont during the Civil War and in the southern Piedmont, Blue Ridge, Ridge and Valley and Appalachian Plateaus during the Depression. The few, small areas of boreal forest that remain probably were never cleared; secondary succession in the spruce forest seems to lead to a forest of northern hardwoods in today's environment.

Clearing of forests means loss of habitat for the species that inhabit them. Clearing usually proceeds in a patchwork fashion that first fragments the forest. For many animals, habitat fragmentation means increased expenditure of time and energy in locating food, increased difficulty in finding mates, and perhaps greater vulnerability to predation when passing through cleared lands. For plants, fragmentation may mean an increase in light and evaporation rates, a decrease in pollinators and dispersers, and/or an increase in predation. Any or all of these factors can cause population decline and increased vulnerability to extirpation.

What may mean decline or extinction for forest species, however, may be a boon to those that require edge or open habitats. White-tail deer, for example, are probably much more numerous in Virginia today than when colonists first arrived (although they first had to be saved from the brink of extinction by reintroductions and enlightened game management practices). Red fox and coyote, both of which prefer open woods, edges, or fields, have moved into Virginia since the Europeanization of the landscape.

Extreme habitat modification accompanies urban development, which frequently destroys natural habitats altogether, replacing them with artificial ones. The new environments are by no means devoid of life, but they generally support neither the same kinds nor the same diversity of species that the former natural habitats did. A distinct fauna and flora develops in towns and cities. Introduced species abound in tended gardens and untended back lots: Japanese honeysuckle, dandelion, pigeon, house sparrow, starling, Norway rat, house mouse. But native species also adapt: blue jay, opossum, raccoon, to name just a few.

Farming, urbanization, and industrialization also have affected streams and estuaries. Our waters are polluted with nutrients and toxins, and siltation rates have increased. The excessive sediments suffocate bottom-dwelling organisms.

In the Appalachian Plateaus Province, drainage from coal mines and from coal storage heaps is highly acidic. Water reacts with the sulfur in the coal to produce sulfuric acid which, when it enters drainage systems, acidifies streams beyond the point at which many forms of life can exist. Siltation from eroded hillsides and from coal fines or dust blowing from trucks and rail cars is also a major problem. The Big Sandy River tributaries are totally devoid of native mollusk species, and the upper Powell River has lost nearly all its native mollusks as a result of siltation and acid mine drainage.

Throughout the Commonwealth, swamps, marshes, and bogs have been drained and converted to more "productive" land uses in human-dominated ecosystems. Drainage of wetlands began early in Virginia's history with a major project focusing on Great Dismal Swamp to reclaim land for agriculture.

Humans have influenced the nature of Virginia for at least 12,000 years. Sometimes we have accelerated natural processes; sometimes we have interrupted them. As a consequence of human activity, species have been lost from Virginia's flora and fauna; but species have also been added. Our landscapes today record the cultural and economic history of Virginia's people as much as they record natural history. Our values and needs and our understanding (or lack of understanding) of how our actions change our environment dictate which other species will continue to share the human habitat with us.

Based on "The Nature of Virginia" by Susan L. Woodward and Richard L. Hoffman, pp. 21–41 in *Virginia's Endangered Species*, coordinated by Karen Terwilliger, 1991 (Blacksburg, VA: The McDonald & Woodward Publishing Company).

Vascular Plants

The term vascular plants refers to those plants having the vascular tissues xylem and phloem that conduct water and nutrients through the plant. All coniferous and broad-leaved trees and shrubs, flowering herbs including grasses and sedges, and ferns and fern allies are included. Algae, lichens, fungi, and mosses and liverworts are nonvascular and are excluded. The latter groups are too poorly known at present in Virginia to enable any being listed as endangered or threatened.

The second edition of the *Atlas of Virginia Flora* lists slightly more that 2,700 species of vascular plants occurring in the Commonwealth. About 2,100 of these are native species. The Division of Natural Heritage monitors about 600 species of vascular plants known or suspected to be rare or vulnerable in the state. The Virginia Department of Agriculture and Consumer Services, the Division of Natural Heritage, the US Fish and Wildlife Service, The Nature Conservancy, and others have undertaken many surveys of Virginia's rare plants and contributed to an extensive data base for assessing their distributions and statuses in this state.

Beyond the few species officially listed as endangered or threatened in Virginia and thereby provided legal protection, many other rare species continue to be monitored to be sure they do not become endangered or threatened, or are not extirpated. The search continues for new populations of the 33 species known only from historical records and those whose status is presently undetermined.

The flora of Virginia is intermediate between that of West Virginia with about 2,200 species and that of the Carolinas with 3,360 species. While species richness generally increases as latitude decreases, for any given region the diversity of species is much more related to diversity of habitats — which is related to diversity of physiography, soil type, climate, and other environmental variables. As is the case for any significantly large region, the Virginia flora contains aggregations of species, termed elements, which have their centers of distribution elsewhere. For example, plant geographers typically speak of the southern element, or more specifically, the southeastern Coastal Plain element, to refer to assemblages of species which characterize the southeastern part of the US and whose distributions in Virginia are somewhat peripheral. Few species associations are uniquely Virginian.

Of the 260 or so species of the southern element in the Virginia flora, most are associated with the Coastal Plain. Some species-rich habitats characteristic of the Coastal Plain south of Virginia are either absent from Virginia or are present in a floristically depauperate form. These include southeastern maritime forests, evergreen shrub pocosins, and turkey oak-longleaf pine savannas. Even so, about 200 species reach their northern limit of distributions in southeastern Virginia.

The Chesapeake Bay and its major tributaries present a major physical barrier to migration for species largely associated with the Coastal Plain.

The Piedmont flora is essentially southern in its affinities. South of Virginia, the Piedmont Plateau continues as a broad expanse between the mountains and the Coastal Plain. Most of its habitats, likewise, are continuous southward for a great a distance. Certain habitats become infrequent and floristically depleted north of the southern border of Virginia. Notable examples of habitats important for rare plants are granite flatrocks, Fall Line sandhills, and diabase glades.

In contrast to the Coastal Plain and Piedmont, the mountain provinces are more nearly allied floristically with the northeastern states; although a strong southern Appalachian element is present through the southern Blue Ridge, especially on Mount Rogers and Whitetop, and the Ridge and Valley Province in southwestern Virginia. In general, the northern element in the flora of Virginia ranges southward throughout most of the highlands; the rarer species of this element are peppered in bogs, fens, spring seeps, and on steep north slopes and windy, rocky summits where conditions are less ameliorated by the austral climate related to Virginia's latitude. The discontinuity of these habitats, their small size, and relative infrequency all contribute to the rarity of many species associated with them.

One pattern of plant geography in Virginia which has received considerable attention is the disjunction of many species occurring both on the Coastal Plain and in the mountains. A number of factors may be involved in this pattern, one being that weathering of sandstone and quartzites in the mountain produces edaphic conditions similar to those found commonly on the Coastal Plain. Such conditions are rare or absent on the Piedmont. In addition, the northward migration of species following the last glaciation advanced as much up the Mississippi Valley as along the Atlantic seaboard. Lowland species migrated along the Tennessee River drainages from the west and along the coast from the south.

The western element in the Virginia flora is primarily associated with the Ridge and Valley Province. The best known group of species with western affinities is associated with areas of Devonian and Ordovician shales on the lower slopes of mountains. These areas, termed shale barrens because of their rocky ground and sparse vegetation, resemble parts of the western United States and support a unique community with many endemic species. Pines and red cedar dominate the canopy much as the pinyon pine and juniper dominate the woodlands of the Great Basin. Plants of the shale barrens are, in some cases, the same species that are found in the West, while others have their nearest relatives there. Although shale barrens are frequent in western Virginia, they are rare elsewhere. The shale barren flora is best developed in western Virginia; two of the three species of *Clematis* found on shale barrens are endemic to Virginia. Although some of these are too common to be considered threatened or endangered, they hold a special place in our natural heritage and deserve our protection.

Several other habitats in the Ridge and Valley Province are important hosts for species characteristic of midwestern prairies. Fens, wet meadows, cedar barrens, and limestone cliffs, often with arbor-vitae, *Thuja occidentalis*, provide suitable conditions for such species. Some species of essentially midwestern distribution are part of the Virginia flora largely because the western extremity of the state is so far west. That all or most of the nine westernmost counties in Virginia are in the Tennessee and Ohio River drainages is an important factor in the occurrence of midwestern species in western Virginia.

Natural factors of physical geography interacting with past biogeographic processes account, in part, for the large number of rare plant species found in Virginia. A number of human-related factors, however, have been and continue to be major threats both to individual species and to their habitats. Severe modification of many natural vegetation types has taken place, usually resulting in the elimination of native plant species and communities. Major factors in the transformation of Virginia's flora are urbanization, loss of wetlands by drainage and filling, deforestation, conversion of forests to pine plantations, and fire suppression. The vegetation of the different physiographic provinces differs in the degree to which it has been affected by these disruptive factors.

Urbanization is a particularly acute threat along the Fall Line and in portions of the Coastal Plain. The proximity of urban areas puts many pressures on nearby natural areas. Certain habitat types that are rare today are mere remnants of what were formerly considerable acreages and were decimated because of the commercial value of products which they sustained. Good examples are the remaining stands of longleaf pine, *Pinus palustris*, and Atlantic white cedar, *Chamaecyparis thyoides*, species highly valued for pine tar and cedar shingles, respectively. These and other communities in the Coastal Plain which are fire adapted have been further degraded by fire suppression. Fire may have played an important ecological role in the mountains as well, but the degree to which suppression has affected the present vegetation of the mountains is uncertain.

A long history of intensive agriculture over the entire Piedmont has been a major factor of change in the vegetation of that province. Even much of the land now forested was cultivated in the past. Areas which have remained continuously forested are some river and creek bluffs too steep to farm and lands with poor agricultural potential due to poor or thin soils or rocky ground. Deforestation for agriculture has been most extensive in the Triassic basins.

Deforestation is also extensive in the Ridge and Valley Province, where broad, flat valleys underlain by carbonate rocks, such as the Great Valley, were desirable sites for farms. In the mountains, timbering, livestock grazing, mining, and quarrying activities all have played roles in changing natural communities.

Loss of wetlands has been a major threat to plant species in all physiographic provinces. Although total acreage of wetlands destroyed has been greatest in the Coastal Plain, the loss of wetlands has perhaps been more devastating in the highlands, where many fewer acres originally existed. Wetland habitats were viewed as unproductive and consequently were drained, filled, or developed for the water they might supply, thereby being so significantly reduced or modified as to become unsuited for many of the species they supported under pristine conditions.

Some species are rare by virtue of political or physiographic boundaries and may be relatively frequent a short distance beyond our borders. Species at or near the edges of their ranges often become very local and may occur in unique or specialized habitats. In Virginia, a number of such peripheral species occur in unique assemblages as exemplified by the co-occurrence of Coastal Plain and northern bog species in sinkhole ponds in the Shenandoah Valley.

The significance of narrow endemics and species of extreme rarity throughout their ranges is intuitively obvious. Interestingly, the Broadleaf Deciduous Forest Formation has given rise to few narrow endemics, and its expanse has given ample opportunity for many species to find suitable sites over a broad geographical area. Thus Virginia's most unusual botanical areas are often defined by such criteria as the presence of disjuncts or relicts or of unique assemblages of species.

The importance of significantly disjunct populations lies in the high probability that they will differ genetically from those populations within the main body of the species' range. Very often disjunct populations are relicts, reflecting the climatic and vegetation history of Virginia, principally since the end of the last major glaciation 10,000 years ago.

The six vascular plant species endemic to Virginia all occur in the mountain region. Virginia round-leaf birch, Addison's leatherflower, Millboro leatherflower, Virginia sneezeweed, and Peters Mountain mallow are all endangered species, while white-haired leatherflower, *Clematis coactilis*, is a species of special concern. All have very narrow distributions, and most occur in restricted habitats as well. That they all occur in the mountain regions may indicate that narrow habitats are more common there than on the Coastal Plain or Piedmont.

Augusta County has far more rare plants that any other county or city in Virginia. It is blessed with such rare habitats as sinkhole ponds, fens, bogs, and shale barrens, which range along with myriad other habitats from the high Alleghenies on the west through the Shenandoah Valley to the high Blue Ridge on the east. Giles, Montgomery, and Grayson counties also have a large number of rare species, the first two because they contain central Appalachian endemics and midwestern disjuncts, and the last because it contains many southern Appalachian endemics. Page and Madison counties have northern montane species that reach Hawksbill and Stony Man mountains. Caroline County has six relict species. Isle of Wight and the City of Suffolk have sandhills and pine barrens with rare species. Five rare species are found in Brunswick County on granite flatrocks.

The interest in rare and endangered species by state, federal, and private organizations has been instrumental in focusing the attention of botanists on not only the rare species themselves, but also on the specialized habitats in which so many occur. Human pressures on natural habitat will continue to grow throughout Virginia. Ability and tenacity in planning the rational uses of vegetation communities and their constituent species are vitally important for the future well being of these communities. It is hoped that the present effort will help direct attention to our more important natural areas and facilitate their protection and preservation.

Abridged from "Vascular Plants" by Duncan M. Porter and Thomas F. Wieboldt, pp. 51–59 in *Virginia's Endangered Species*, coordinated by Karen Terwilliger, 1991 (Blacksburg, VA: The McDonald & Woodward Publishing Company).

Variable Sedge

Carex polymorpha

ENDANGERED

DESCRIPTION: Variable sedge has stout, erect culms, 12 to 24 inches high, that emanate from cord-like creeping rhizomes. Narrow leaves are shorter than the culm. Staminate spike is usually solitary; pistillate spikes 1 or 2, erect, ⁶/₁₀ to 1½ inches long. DISTRIBUTION: In Virginia, this species is known from only eight populations, some numbering scores of plants, on sandstone tablelands in the Allegheny Mountains and a granitic summit in the Blue Ridge Mountains. The station in Alleghany County, at an elevation of about 3,400 feet above level, is the southernmost occurrence in its known range. HABITAT: This sedge has been found almost entirely on acidic, sterile, mountain top soils underlain by resistant siliceous sandstones of the Clinch formation. One popu-

lation is on similarly poor and acidic soils derived from granitic rocks of the Pedlar formation. Occurring at elevations of 3,100 to 4,300 feet above sea level, variable sedge forms colonies in sandy openings and road banks in oak-pine-heath communities and thin dry oak woods. All locations have a long history of fire. LIFE HISTORY: Colonies of variable sedge are largely vegetative. New shoots are derived from its stout, cord-like rhizome. Flowering is from June through August but occurs infrequently and usually where some disturbance has opened the forest canopy. THREATS: It would appear that human activities, such as clearing of vegetation and road building, might actually benefit this species. Most stations are in the George Washington National Forest or Shenandoah National Park. NOTES: The recent discovery of new populations of variable sedge in Virginia is attributed to botanists having learned to recognize the species in its vegetative condition in areas where it formerly went unnoticed.

Abridged from "Variable Sedge" by Charles E. Stevens, pp. 73–74 in *Virginia's Endangered Species*, coordinated by Karen Terwilliger, 1991 (Blacksburg, VA: The McDonald & Woodward Publishing Company).

Harper's Fimbristylis

Fimbristylis perpusilla

ENDANGERED

DESCRIPTION: Harper's fimbristylis is a diminutive, solitary or tufted annual sedge. Culms may be up to 6 inches long, but normally are much shorter. Each clump displays culms of varying lengths, shortest at ground level. The inflorescence is a compound umbel, doubly so on robust individuals. Spikes narrowly ovoid, greenish to pale brown, sometimes suffused with tawny coloration. The central spike is sessile and subtended by several leaflike bracts exceeding the length of the inflorescence. Attenuated and slightly recurved scales lend a bristly appearance. Achene is banana-shaped, about 1/64 inch long. Among species of *Fimbristylis* in the United States, this species is readily distinguished by its two style branches and its achene shape. DISTRIBUTION: Harper's fimbristylis is known in Virginia only from near Grafton, York County. HABITAT: In Virginia, this species is known from ephemeral, summer-dry ponds, where it tends to grow in the pond bottoms and away from other plants. Ponds created during periods of high moisture and low evapotranspiration provide moist, silty, and sunlit conditions as the waters gradually recede during the growing season. LIFE HISTORY: Harper's fimbristylis is an annual which appears from mid-summer to late fall. The species is apparently capable of long periods of dormancy, being known to virtually disappear for years. A seedbed of bare mud free of dead leaves and other detritus seems to be required. THREATS: The primary threats to this small sedge come from the potential for major disturbance to its habitat, filling and drainage being the most obvious. A more subtle form of drainage would be drawdown of the water table. Vernal ponds are also victim to the increasing use of recreational vehicles. NOTES: In July 1986, Harper's fimbristylis was first discovered in Virginia occupying about 20 square feet of one of the Grafton ponds. Subsequent efforts to find this or other occurrences by the staff of the Division of Natural Heritage resulted in the discovery in 1991 of a second population in a nearby vernal pond. The species was listed in Virginia as endangered in 1989.

Abridged from "Harper's Fimbristylis" by Thomas F. Wieboldt, pp. 79–80 in *Virginia's Endangered Species*, coordinated by Karen Terwilliger, 1991 (Blacksburg, VA: The McDonald & Woodward Publishing Company).

Northeastern Bulrush

Scirpus ancistrochaetus

ENDANGERED

DESCRIPTION: Northeastern bulrush has culms 32 to 48 inches high, sometimes proliferating from upper nodes. The inflorescence of divergent and arching rays bears clusters of brown spikelets; scales are brown or blackish with green midrib. Achenes are obovoid, light yellow brown, tiny, subtended by 6 bristles about the same size as the achene and downwardly barbed nearly to the base with rigid, thick-walled teeth. Northeastern bulrush closely resembles *Scirpus atrovirens* in size and habit and is distinguished from it primarily by the latter's slightly smaller achenes, which have a variable number of bristles delicately barbed in the distal half or only near the tips. DISTRIBUTION: Northeastern bulrush is extremely rare in Virginia, where it is known from only six mountain ponds. These are the southernmost records for the species. HABITAT: In Virginia, northeastern bulrush is part of a relict boreal flora associated with sinkhole and mountain ponds of the Ridge and Valley Province. Sites are often characterized as bogs with pH varying from acidic to slightly alkaline. The species is predominantly associated with solution basins where acidic alluvium or colluvium overlies limestones and dolomites. In Virginia, this bulrush occupies the deeper parts of the emergent zone. LIFE HISTORY: Northeastern bulrush is a perennial; it flowers from mid-June to July. Heads remain in relatively good condition through the remainder of the summer. Fruits mature from July to September. Most details of the life history of this species are unknown. THREATS: Since this bulrush occupies very wet sites, any modification which lowers the water level or dries out the habitat could eliminate all or most individuals in any of the populations. Siltation and competition from introduced weedy species could become a threat if habitats surrounding the ponds are not protected. Under disturbed conditions, introgressive hybridization could become a problem. NOTES: This species was not described until 1962. The mountain ponds where this species is found are about 11,000 years old and harbor some of the older elements of Virginia's modern vegetation. The co-occurrence of other species not generally distributed in the region lends credence to the idea that the northeastern bulrush is a relict species. This species was listed as endangered in Virginia in 1989.

Abridged from "Northeastern Bulrush" by Thomas F. Wieboldt, pp. 80–82 in *Virginia's Endangered Species*, coordinated by Karen Terwilliger, 1991 (Blacksburg, VA: The McDonald & Woodward Publishing Company).

Swamp Pink

Helonias bullata

ENDANGERED

DESCRIPTION: Swamp pink is a smooth perennial herb growing to 3 feet high. Leaves are basal, oblanceolate, and evergreen and grow from a hard tuberous rhizome. Fragrant flowers form a dense raceme; the perianth is in six segments, showy, lilac pink; anthers are silver blue. DISTRIBUTION: In Virginia, swamp pink is known from about 30 stations, two on the Coastal Plain and the rest within about a five mile radius in and at the foot of the Blue Ridge in Augusta and Nelson counties. HABITAT: Swamp pink is found in water-saturated, usually organic soil or black muck which is mostly sphagnous. These habitats are perma-

nently wet locations which are distinctly acidic. The species does not tolerate inundation; consequently it is only located along small waterways, in springy ground, or other areas where water levels are stable and not subject to flooding except very infrequently or briefly. Although tolerant of sun, it is usually found in shaded habitats. LIFE HISTORY: Population maintenance is predominantly by vegetative production of new rosettes rather than by seed. THREATS: The Henrico County colony, which had about 1,000 plants in 1987, is situated in woods beside a railroad, but is probably in no danger at present. Most of the Blue Ridge stations lie in the George Washington National Forest and are probably subject to damage only from logging and road building. NOTES: *Helonias bullata* is the only member of the genus. Its closest relative is the Asian genus *Heloniopsis*. The swamp pink is federally listed as threatened, but in Virginia it is listed as endangered.

Abridged from "Swamp Pink" by Charles E. Stevens, pp. 88–89 in *Virginia's Endangered Species*, coordinated by Karen Terwilliger, 1991 (Blacksburg, VA: The McDonald & Woodward Publishing Company).

Prairie White Fringed Orchid

Habenaria leucophaea

THREATENED

DESCRIPTION: Prairie white fringed orchid is a perennial with stems 36 to 40 inches high and large showy racemes, 3 to 8 inches long and 2 to 3 inches in diameter. Flowers are creamy white; lateral petals are broadly obovate, lip deeply three-parted, spur 1 to 1½ inches long. In Virginia, it is most similar to the normally pink- or magenta-flowered large purple orchid, *Habenaria fimbriata*, and small purple fringed orchid, *Habenaria psychodes*. Among other white-flowered species, white fringed orchid, *Habenaria blephariglottis*, lacks the divided lip and is whiter; ragged fringed orchid, *Habenaria lacera*, has much smaller flowers which are greenish in color. DISTRIBUTION: This orchid is known in Virginia from a single site along South River in Augusta County. HABITAT: In Virginia, prairie white fringed orchid occurs is an open, pastured wet meadow. LIFE HISTORY: This orchid is believed to be a long-lived perennial tolerant of disturbance and capable of prolonged dormancy and responding quickly and positively to fire. Rapid population fluctuations are characteristic in other parts of its range. Flowers remain open for about 10 days, providing for simultaneous flowering of much of the inflorescence. The species is believed to be adapted for pollination by hawkmoths by its nocturnal fragrance, white color, and long spurs holding large amounts of nectar. THREATS: Cattle grazing and trampling at the Augusta County site have hindered development and dissemination of seed, but alternatively, have helped to control competing vegetation which may preclude germination or establishment of seedlings. Recent ponding by beaver dams may have adversely affected the population. Long term threats include shading, were grazing or haying to cease, and hydrologic changes resulting from nearby industrial development. NOTES: Prairie white fringed orchid was discovered in Virginia in 1979. Its occurrence in Augusta County suggests that presettlement vegetation may have provided prairie or prairie-like habitats not available elsewhere in the region. If it is a relict, some of the threats it faces today are not new. The degree to which human manipulation of the habitat can substitute for natural disturbances is crucial to proper management. Species of *Habenaria* in Virginia are placed in the genus *Platanthera* by some botanists. An alternative common name for this species is eastern prairie fringed orchid to distinguish it from its western relative. Prairie white fringed orchid is federally listed as threatened.

Abridged from "Prairie White Fringed Orchid" by Thomas F. Weiboldt, pp. 93–95 in *Virginia Endangered Species*, coordinated by Karen Terwilliger, 1991 (Blacksburg, VA: The McDonald & Woodward Publishing Company).

Small Whorled Pogonia

Isotria medeoloides

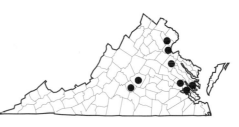

ENDANGERED

DESCRIPTION: Small whorled pogonia is a perennial. Flowering plants are 4 to 10 inches high; vegetative plants are shorter. Stems are robust, hollow, smooth, pale green, and glaucous — like white grapes. Leaves pale green, glaucous, and borne in a single whorl of 5 or 6 at the top of the stem. The whorl reflexes prior to flowering, then becomes horizontal. One or two flowers form in the center of the whorl. Sepals as long as or slightly longer than petals. Large whorled pogonia, *Isotria verticillata*, has a reddish purple stem and dark green, non-glaucous stems. Its sepals are purplish brown and two to three times as long as the petals. DISTRIBUTION: In Virginia, small whorled pogonia has been documented from seven counties on the Piedmont and Coastal Plain. Extant colonies are known from only five counties. HABITAT: In Virginia, small whorled pogonia occurs in very ordinary looking third growth upland forests on terrain that is almost level or gently to moderately sloping in northerly or easterly directions. The understory is distinctly open, and flecks of sunlight play on the forest floor throughout the day. Some and perhaps all of the colonies occur on land that has been previously cultivated. Soils are acidic sandy loams with low to very low nutrient contents by agricultural standards. LIFE HISTORY: Flowering typically begins in late April to mid-May. Flowering is so well synchronized that the total flowering period within a colony occurs within two and one-half weeks. Some colonies are composed mostly of vegetative plants, others mostly of flowering plants. Small whorled pogonia is self-pollinated and rarely produces more than one stem per plant. A plant which produces a large, flowering stem one year may appear the next year as a diminutive vegetative plant. THREATS: Populations of small whorled pogonia in Virginia are particularly threatened by the development of housing subdivisions. Large populations of deer also are a threat because the plant usually does not reappear the next year when its whorl is grazed early in the season. Some colonies have survived selective timbering, but clearcutting and other practices resulting in drastic changes in light factors or significant increase in interspecific competition would likely cause a colony to decline. NOTES: Small whorled pogonia was listed as a federal endangered species in 1982 and as endangered in Virginia in 1985.

Abridged from "Small-Whorled Pogonia" by Donna M. E. Ware, pp. 95–97 in *Virginia's Endangered Species*, coordinated by Karen Terwilliger, 1991 (Blacksburg, VA: The McDonald & Woodward Publishing Company).

Virginia Round-Leaf Birch

Betula uber

ENDANGERED

DESCRIPTION: Virginia round-leaf birch is a deciduous, single-trunked tree that reaches a height of about 50 feet. Branches are spreading to ascending, forming a rounded to oval crown. Bark is dark brownish black, dull to glossy, smooth to slightly roughened with horizontal lenticels, and has a wintergreen aroma when bruised. Twigs are reddish to dark brown, lenticelate. Leaves simple, alternate, ovate to nearly round, ¾ to 1¾ inches wide, ¾ to 2¼ inches long, with three to five pairs of lateral veins, coarsely serrate, rounded at apex and heart-shaped at base. Among the birches, this species can easily be distinguished by its tree-like habit; aromatic, smooth, blackish bark; and nearly round leaves with only 3–5 pairs of lateral veins. It differs from sweet birch, *Betula lenta*, primarily in leaf shape; the latter having leaves that are elliptical to ovate, pointed apex, and with eight or more pairs of lateral veins. The two species grow together at the single known locality of native round-leaf birch. DISTRIBUTION: Virginia round-leaf birch is represented by only one indigenous population along Cressy Creek in Smyth County, Virginia. Several test populations have been established for research purposes. HABITAT: The single known indigenous population is scattered within a mixed, open secondary forest. It is a subcanopy tree. It grows in a stony colluvium that is strongly acidic and very permeable. LIFE HISTORY: Virginia round-leaf birch flowers in late April to early May. Fruits mature the following September. Reproduction is presumably sexual. THREATS: The primary threat to the indigenous population and the several research plots is vandalism (human and animal) even though most of the natural population is on land now owned by The Nature Conservancy. NOTES: Herbarium specimens of Virginia round-leaf birch were first collected in 1914. In 1954 it was reported as no longer existing along Dickey Creek, the only published location for the species. It was ultimately rediscovered along Cressy Creek in 1975. Extensive searches of the area have not revealed any other natural populations. Virginia round-leaf birch was the first plant to be federally listed as endangered. It was listed as endangered in Virginia in 1979. An upgrade in status to threatened is currently proposed at both state and federal levels.

Abridged from "Virginia Round-Leaf Birch" by Peter M. Mazzeo, pp. 97–99 in *Virginia's Endangered Species*, coordinated by Karen Terwilliger (Blacksburg, VA: The McDonald & Woodward Publishing Company).

Buckleya

Buckleya distichophylla

ENDANGERED

DESCRIPTION: Buckleyas are dioecious shrubs up to 15 feet high. Stems
are much branched, green and puberu-
lent when young, brown and smooth
when older. Scattered white lenticels appear on older stems. Leaves are opposite. Short
lateral branches give the superficial appearance of compound leaves. Leaf blades are lanceolate, ½ to 2 inches long and ¼ to ¾ inches wide. Staminate inflorescence umbel-like,
erect, terminal; staminate flowers rotate and green. Pistillate flowers single, terminal, pen
dulous, and green with four deciduous sepals. Buckleya differs from the related nestronia
in height and flower structure. Nestronia is only 20 inches high with staminate inflorescences in leaf axils. DISTRIBUTION: In Virginia, buckleya is known from seven counties
in the southwestern mountains. HABITAT: Buckleya favors shaly, often very steep slopes,
frequently along streams with a westerly exposure. LIFE HISTORY: Buckleya is a root
parasite. Flowering takes place in May. Small flies are recorded as visiting the flowers, but
little else is known about the species' floral biology or breeding patterns. THREATS: The
major threat to buckleya in Virginia is road
construction, which has already drastically reduced populations in Craig and Montgomery
counties. The acquisition by The Nature Conservancy of the large population in Roanoke
County ensures the perpetuation of the species in the state, but care is needed to preserve
as much genetic diversity as possible. Buckleya
has sometimes been over-collected by zealous
botanists eager to have this rare shrub in
herbariums or to distribute as an exchange
specimen. An additional threat is browsing of
young shoots, presumably by deer. NOTES:
Buckleya distichophylla is well known to North
American botanists as one of the rarest shrubs
on the continent. The genus consists of only
two species, the other occurring in Japan and
China. Buckleya is known from the Appalachian Mountains of Tennessee, North Carolina, and Virginia, where the largest
populations are found. The species was listed
as endangered in Virginia in 1989.

Abridged from "Buckleya" by Lytton J. Musselman, pp. 99–100 in *Virginia's Endangered Species*, coordinated by Karen Terwilliger,
1991 (Blacksburg, VA: The McDonald & Woodward Publishing Company).

Nestronia

Nestronia umbellula

ENDANGERED

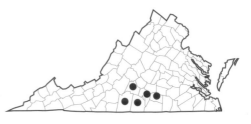

DESCRIPTION: Nestronia is a deciduous, rhizomatous shrub. Stems are dark
brown and up to 20 inches high. Leaves
1¼ to 3⅜ inches long. Staminate inflo-
rescences on erect axillary umbels with 5 to 10 flowers; staminate flowers ⅛ in long, and
1/10 inch wide. The three to five sepals are reflexed, green, and fragrant. Pistillate flowers
are unknown in Virginia. Nestronia differs from buckleya in that the latter is a shrub
growing over 15 feet high and has terminal staminate inflorescences and rotate staminate
flowers. DISTRIBUTION: In Virginia, this species is known from the southern Piedmont. HABITAT: The largest populations are
along a narrow stream in Pittsylvania County,
where the plants are growing in the sandy margins and on adjacent slopes. In Halifax County,
nestronia is restricted to the slopes adjacent to
a creek. The site is moderately steep, well-
drained, sandy, and shady. Nestronia forms a
clone which excludes all but a few herbaceous
species. There are indications that nestronia is
an ecotonal plant favored by burning or other
disturbance. LIFE HISTORY: Unfortunately
we know very little of this shrub other than
that it is a root parasite. Nestronia has been
found to parasitize a variety of hosts. In Virginia, all reproduction is asexual by rhizomes.
Staminate plants flower in May. THREATS:
The greatest potential danger to nestronia is
habitat destruction. NOTES: *Nestronia* is a
monotypic genus and occurs in Virginia, North
Carolina, South Carolina, Georgia, Alabama,
and Tennessee. It was listed as endangered in
Virginia in 1989.

Abridged from "Nestronia" by Lytton J. Musselman, pp. 100–101 in *Virginia's Endangered Species*, coordinated by Karen Terwilliger,
1991 (Blacksburg, VA: The McDonald & Woodward Publishing Company).

Addison's Leatherflower

Clematis addisonii

ENDANGERED (proposed)

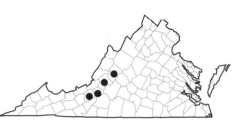

DESCRIPTION: Addison's leatherflower is a perennial, semi-shrubby vine from a rather stout, short underground caudex. Young stems are erect; by fruiting time they are often arching, leaning, or scrambling. Stems may be either simple or branching from upper nodes and up to 3 feet long. Stems die back to caudex at end of growing season. Leaves opposite and quite variable in shape and size. Flowers are usually nodding, arising singly or doubly from stem tips or oppositely from axils of leaves. Sepal back usually glabrous, but tomentose on margin distally. This *Clematis* is one of the rarest and most elusive species in the genus. It is not likely to be confused with any other species of *Clematis* in Virginia. **DISTRIBUTION:** Addison's leatherflower is endemic to the Ridge and Valley Province of Virginia. It was first collected in 1881, but only 14 sites are known. This species no longer exists in some of the old sites where it was once collected. Most individuals occur in Montgomery County. **HABITAT:** Addison's leatherflower is restricted to rocky, dry limestone hillsides, wooded bluffs, banks, ledges, and ravines. Sometimes it occurs on relatively dry shaly soils over limestone or dolomite. On thin soils, only scattered individuals occur. Plants may be in areas that are open or lightly shaded in mixed stands of hardwoods and red cedar. Preferred soils appear to be well-drained or porous. **LIFE HISTORY:** Several years of drought reduce the numbers in populations severely. Populations are large only at Falls Ridge, Montgomery County. Flowering is from May into early July. **THREATS:** Several of the largest populations are on the banks of county roads. If the roads were ever widened these sites would be destroyed. **NOTES:** Addison's leatherflower is protected in The Nature Conservancy's Falls Ridge Nature Preserve in Montgomery County.

Abridged from "Addison's Leatherflower" by Gwynn W. Ramsey, pp. 106–107 in *Virginia's Endangered Species*, coordinated by Karen Terwilliger, 1991 (Blacksburg, VA: The McDonald & Woodward Publishing Company).

Millboro Leatherflower

Clematis viticaulis

ENDANGERED (proposed)

DESCRIPTION: Millboro leatherflower is an herbaceous perennial with stems clustered or clumped from a woody rhizome. Stems are erect or nearly so, reddish brown, and 12 to 20 inches high. Leaves opposite, simple, ovate or lance ovate, 2 to 3¼ inches long, sessile. Flowers are about ¾ inch long, the four sepals tinted blue and maroon. Reddish brown achenes are a distinctive feature. Millboro leatherflower differs from curly heads, *Clematis ochroleuca*, in its narrower leaves, sepals only minutely pubescent externally, and styles only ¾ to 1¼ inches long at maturity and covered with deep brown hairs. It is distinguished from white-haired leatherflower, *Clematis albicoma*, by its canescent sepal backs, hirtellous stem pubescence, and reddish brown achene tails. DISTRIBUTION: Millboro leatherflower is endemic to Virginia, found only in Bath and Rockbridge counties. This species grows only on shale barrens. There are only about ten sites, and most colonies are small. HABITAT: Millboro leatherflower is a strict shale barren endemic. The plants grow on steep southward facing slopes up to at least 50°. Individuals may be found on rather open barrens or in partial shade. LIFE HISTORY: Millboro leatherflower flowers from May through June or early July and tends to grow in clumps from one rhizome. Fruits are found from June through September. It may produce both asexually by rhizome growth and sexually by seeds. THREATS: Threats from humans are minimal. However, in some areas, bulldozing for land fill material is a definite possibility. Fire could be a natural hazard, as could be fracturing and slippage of the thin-bedded shales. Shading by tree growth may eliminate this species, since its habitat is usually very open and dry. NOTES: This is one of the most beautiful species of *Clematis*. Because of its rarity and the rarity of sites, it is imperative to acquire all or a portion of a shale barren supporting this species to insure its preservation and protection.

Abridged from "Millboro Leatherflower" by Gwynn W. Ramsey, pp. 107–108 in *Virginia's Endangered Species*, coordinated by Karen Terwilliger, 1991 (Blacksburg, VA: The McDonald & Woodward Publishing Company).

Shale Barren Rockcress

Arabis serotina

ENDANGERED

DESCRIPTION: Shale barren rockcress is a biennial herb, 16 to 40 inches high. Stems erect, solitary, branched above and glabrous. Rosette leaves are spatulate and toothed and often withered by flowering time. Flowers are small with white or creamy petals and clustered in racemes. Shale barren rockcress has long been confused with Burk's smooth rockcress, *Arabis laevigata* var. *burkii*, a spring-flowering plant of similar habitats and similar leaf morphology, but which is smaller in stature and has larger flowers. Shale barren rockcress blooms in the summer. Sicklepod, *Arabis canadensis*, is occasionally misidentified as shale barren rockcress because of its similar flowering period, but differs by having simple racemes, pubescent leaves, and sparingly pubescent stems. DISTRIBUTION: Shale barren rockcress is an endemic of mid-Appalachian shale barrens. In Virginia, it is known from 16 sites in the Ridge and Valley Province, including the Massanutten Mountains. It seems to be absent from both the northernmost and southernmost barrens in the state. HABITAT: Shale barren rockcress is a strict endemic of shale barrens. Most individuals are found in partial shade, often near the borders of barrens or near the bases of trees. Populations range in aspect from southeast to west and in elevation from 1,300 to 2,700 feet above sea level. LIFE HISTORY: Shale barren rockcress is a facultative biennial. Flowering occurs from mid-June to September. Populations generally contain only a few to several dozen plants. Numbers of flowering plants in populations fluctuate radically from year to year. This species demonstrates a great capacity for vegetative reproduction. THREATS: The greatest threat to this species appears to be browsing by deer. However, damage attributed to deer may in fact be insect damage. Season-long drought can account for large scale reproductive failure. NOTES: Extensive areas of available habitat in Virginia have not been surveyed for its presence, and it is highly likely that additional sites will be found. This species was listed as federally endangered in 1989, and the same year was listed as endangered in Virginia.

Abridged from "Shale Barren Rockcress" by Thomas F. Wieboldt, pp. 108–110 in *Virginia's Endangered Species*, coordinated by Karen Terwilliger, 1991 (Blacksburg, VA: The McDonald & Woodward Publishing Company).

Small-Anthered Bittercress

Cardamine micranthera

ENDANGERED

DESCRIPTION: Small-anthered bitter- cress is a relatively small (to about 15 inches), erect, white-flowered member of the Mustard Family. As the name implies, a diagnostic feature is its small anthers, which are only about ¹/₅₀ of an inch long. Its leaves bear one pair of lateral leaflets (occasionally a second) below the primary termi- nal one, which tapers to the leafstalk and has irregular and shallow lobes or points along the margin. The species most closely resembles round-leafed bittercress, *Cardamine rotundifolia*, which differs in having somewhat lax stems with long, proliferating branches from the upper stem that trail across the ground. It also has broad-based, even heart- shaped, primary leaflets and anthers about three times as long as those of small-antered bittercress. **DISTRIBUTION:** Endemic to Virginia and North Carolina, small-anthered

bittercress is known in Virginia only from Patrick County. Populations are presently known from Peter's Creek and Russell Creek and their tributaries. **HABITAT:** Small- anthered bittercress occupies a diversity of wooded, wetland habitats ranging from sand and gravel bars, stream banks, spring branches, and seepage areas to seasonally wet floodplain woods along small streams. **THREATS:** The principal threats are siltation and loss of forest canopy. Removal of the forest cover can severely degrade habitats by opening the way for aggressive, weedy species to occupy and dominate the sites. Conversion of land to ag- ricultural and residential development can destroy habitat altogether. **NOTES:** This species was not discovered in Virginia until 1990. Additional populations have since been found but its entire distribution is imperfectly known. It should be looked for along streams beyond its presently known range. Small- anthered bittercress was listed as a federally endangered species in September 1988.

Prepared by Thomas F. Wieboldt, Massey Herbarium, Virginia Polytechnic Institute and State University, October 1993.

Virginia Spiraea

Spiraea virginiana

ENDANGERED

DESCRIPTION: Virginia spiraea is a clonal shrub 3 to 10 feet high. Mature stems are dark gray, often arching or nearly horizontal; young stems are upright and greenish yellow to reddish brown. Leaves are alternate and very variable in shape, size, and degree of serration. Flowers bright to creamy white in tightly packed corymbs, ranging in size from 2 to 9 inches wide. Corymbs in late summer and fall often dry and persist during winter, making preliminary field identification possible most of the year. **DISTRIBUTION:** Virginia spiraea is endemic to the Appalachians. All localities are in the Appalachian Plateaus or the southern Blue Ridge physiographic provinces. In Virginia, it is represented by fewer than 20 clones in three drainages. **HABITAT:** This plant has a very specific niche. It occurs along scoured banks of high gradient streams or on meander scrolls, point bars, natural levees, and braided features of lower stream reaches. It may be associated with many other riverine species. **LIFE HISTORY:** Seed production seems sporadic, and seedlings have never been observed in the wild. Virginia spiraea is propagated from the horizontal rootstock; natural layering often produces a dense clone that may later fragment and establish down-

stream. Fragmentation by erosion or scour and subsequent downstream dispersal may be the most important means of spreading clones. The most important factor in maintaining a clone seems to be removal of woody competition by erosion. Scour must be sufficient to remove woody trees and vines without washing out the horizontal root. **THREATS:** Certainly the most obvious threat to Virginia spiraea is dam building that would inundate habitat, stabilize downstream water flow, and allow woody competition to overcome clones. **NOTES:** The plant is often difficult to see enmeshed with competitors, and likely habitats are often difficult to search. Many miles of likely habitat have been searched without success. Even so, Virginia has significant populations of this species when compared to other states in its range. Virginia spiraea was federally listed as threatened in 1990. It was first discovered in Virginia in 1985 and listed as endangered here in 1989.

Abridged from "Virginia Spiraea" by Doug Ogle, pp. 117–119 in *Virginia's Endangered Species*, coordinated by Karen Terwilliger, 1991 (Blacksburg, VA: The McDonald & Woodward Publishing Company).

Northern Joint-Vetch

Aeschynomene virginica

ENDANGERED (proposed)

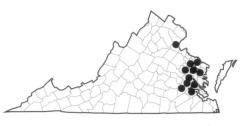

DESCRIPTION: Northern joint-vetch is a robust, bristly-stemmed annual. Leaves are even-pinnate. Flowers yellow, red-veined, and pea-shaped. Two other joint-vetches have been confused with northern joint-vetch. *Aeschynomene indica* and *Aeschynomene rudis* — the former a widespread Old World species, the latter a species centered in the New World subtropics — have both spread northward as far as North Carolina. **DISTRIBUTION:** In Virginia, only eight sites are known. The total size of the population for Virginia is estimated at 700 individuals. Only one population located on the Rappahannock River is considered large enough to maintain itself adequately by natural reproduction **HABITAT:** Northern joint-vetch occurs in high-diversity, slightly brackish tidal marshes of river shores and river banks in a zone generally dominated by annual species. The substrate may be sandy, muddy, or gravelly. **LIFE HISTORY:** Seeds of northern joint-vetch germinate by early June and grow quickly. The plants reach 6 to 18 inches in height by mid-summer, prior to the onset of flowering, which begins about the first of August. Small bumblebees seen visiting the flowers may be pollinators. Flowering continues into late October. The legumes break into one-seeded segments and are dispersed by flotation. Stands of joint-vetch may reappear many consecutive years at isolated sites. Yet other colonies have been noted to exhibit radical population changes from year to year. Storms or other natural disturbances that disrupt vegetation cover on the shore may enhance establishment. **THREATS:** Threats to northern joint-vetch in Virginia include any activity that destroys marshland or results in loss of diversity of species within the marsh. Such threats include bulkheading, damming tributaries, dredging, landfilling, riprapping, and construction activities that directly affect the marsh shore or result in siltation and general water pollution. Deposits of flotsam or refuse could inhibit germination and thereby result in the diminution of populations. Invasion by reed, *Phragmites australis*, has been observed in Maryland and could become a serious threat in Virginia.

Abridged from "Northern Joint-Vetch" by Donna M. E. Ware, pp. 119–121 in *Virginia's Endangered Species*, coordinated by Karen Terwilliger, 1991 (Blacksburg, VA: The McDonald & Woodward Publishing Company).

Running Glade Clover

Trifolium calcaricum

ENDANGERED (proposed)

DESCRIPTION: Running glade clover is a small herbaceous plant with typical three-parted clover-type leaves with elliptical leaflets and long, trailing stems (stolons). Its creamy white flowers have wine-colored veins and occur in round clusters on short, leafless stalks at the tips of the stolons. Flowers strongly resemble those of Kate's Mountain clover, *Trifolium virginicum*, which has long narrow leaflets and lacks the long spreading stems of the running glade clover. The species most closely resembles running buffalo clover, *Trifolium stoloniferum*, a species not yet known from Virginia and which differs in bearing its flower heads from leaf-bearing branches from the base of the plant or from leaf axils along its stolons and never at its tip as in running glade clover. Although not so closely related, running glade clover also resembles the common white lawn clover, *Trifolium repens*, but that species also produces its flowers along the stolons rather than at their tips, and its flowers lack the wine-colored veins. DISTRIBUTION: Running glade clover is found at only two locations: The Cedars in Lee County, Virginia, and a site nearly 200 miles away in Tennessee. Within The Cedars, 24 population centers have been identified over a five square mile area. HABITAT: This species is restricted at The Cedars to rocky, dry limestone woodlands dominated by the red cedar which gives the area its name. The Cedars' scrubby woodlands are a result of stressful conditions created by shallow, nutrient poor soils derived from Hurricane Bridge limestone. The running glade clover is usually found in thinly wooded to open places within The Cedars, although plants can survive in more deeply shaded areas. LIFE HISTORY: Running glade cover "runs" along the ground by rooting at the tips of horizontal stolons which emanate from the base of an existing rooted plant. In addition to such asexual reproduction, this plant also reproduces sexually. Flowering is in May and usually is heaviest in plants which are in partial to full sun. THREATS: At The Cedars, running glade clover is threatened by the proposed development of an airport and a state prison. Secondary impacts, such as road improvements and adjunct commercial development, could also threaten the species. A portion of The Cedars population has been protected through acquisition of a small land parcel by The Nature Conservancy. NOTES: *Trifolium calcaricum* was described as a new species of clover in 1992, adding yet another species to the list of so-called cedar glade endemics. It should be looked for in other barren limestone landscapes in southwestern Virginia.

Prepared by Thomas F. Wieboldt, Massey Herbarium, Virginia Polytechnic Institute and State University, November 1993.

Long-Stalked Holly

Ilex collina

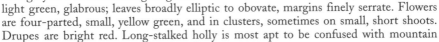

ENDANGERED

DESCRIPTION: Long-stalked holly is a deciduous dioecious shrub 10 to 15 feet high. Branches are spreading and bark gray and smooth. Young twigs are light green, glabrous; leaves broadly elliptic to obovate, margins finely serrate. Flowers are four-parted, small, yellow green, and in clusters, sometimes on small, short shoots. Drupes are bright red. Long-stalked holly is most apt to be confused with mountain holly, *Ilex montana*, which has white flowers and a corolla less deeply lobed. DISTRIBUTION: Long-stalked holly is endemic to the southern Appalachian Mountains. It is known from four counties in southwestern Virginia. HABITAT: Long-stalked holly is confined to a very specialized habitat: high elevation wetlands. It grows at elevations in excess of 3,000 feet above sea level, in bogs and along streamheads in open or shrubby areas. It is also found along medium to large streams at high elevations — an infrequent combination in Virginia. LIFE HISTORY: Long-stalked holly reproduces sexually. Flowering is in May, and fruiting is from June through September. THREATS: The greatest potential threat to the continued existence of long-stalked holly in Virginia appears to be disturbance resulting from logging. NOTES: In many manuals long-stalked holly is treated as Georgia holly, *Ilex longipes*, a more southern species that does not occur in Virginia.

Abridged from "Long-Stalked Holly" by Duncan M. Porter, pp. 126–127 in *Virginia's Endangered Species*, coordinated by Karen Terwilliger, 1991 (Blacksburg, VA: The McDonald & Woodward Publishing Company).

Michaux's Sumac

Rhus michauxii

ENDANGERED

DESCRIPTION: A small deciduous shrub 1 to 3 feet tall, colonies develop from creeping rootstocks. The entire plant is densely pubescent. Leaves have 9 to 13 stalkless, ovate to oblong, coarsely toothed leaflets. Male and female flowers are borne on separate plants in June and July, with female plants subsequently bearing dense, erect clusters of fleshy red fruits. Michaux's sumac is distinguished from other southeastern species of sumac by its short stature, dense overall pubescence, and relatively broad, sharply toothed leaflets. DISTRIBUTION: The single known Virginia population of Michaux's sumac is located on the Fort Pickett Military Reservation in the Piedmont region of Nottoway and Dinwiddie counties. HABITAT: The population is confined to a 10,000 acre artillery training area which is subject to frequent and intense fires from incendiary ammunition. Within this area, colonies are widely scattered in thinly canopied oak-hickory woodlands, grassy hardwood savannas, openings in sprout thickets, and old clearings. THREATS: Shading of habitats from fire suppression and resultant secondary succession is the primary threat; however, the site's current land use minimizes this threat. NOTES: Historically endemic to the inner Coastal Plain and Piedmont from Georgia to North Carolina, Michaux's sumac was discovered in Virginia in 1993. Only 25 extant occurrences of this diminutive shrub are currently known, and the species is federally listed as endangered. The Virginia population at Fort Pickett is not only the northernmost, but also the largest of the species, containing an estimated 20,000 individual stems. the factors contributing to this exceptional status are not entirely understood, but probably relate to the long-term, fire-maintained habitat.

Prepared by Gary P. Fleming, Virginia Department of Conservation and Recreation, Division of Natural Heritage, November 1993.

Peters Mountain Mallow

Iliamna corei

ENDANGERED

DESCRIPTION: Peters Mountain mallow is a perennial herb from a woody rhizome. The stems are erect, sparsely-branched, pale green, densely stellate-pubescent, and up to 40 inches tall. Leaves are simple, palmate, and stellate-pubescent. Flowers solitary or clustered in axils of upper leaves, odorless, to 2 inches in diameter. Petals are pink, spatulate, notched, basally pubescent with long simple hairs 1 to 1¼ inch long. The species has long been confused with the Kanakee mallow, *Iliamna remota*, which differs by growing over 6 feet tall, having leaves with broadly triangular lobes, and having fragrant flowers occurring in long terminal racemes. DISTRIBUTION: Peters Mountain mallow is endemic to Giles County, where it is found in a single population on Peters Mountain above the New River at The Narrows. HABITAT: This rare species grows in pockets of shallow soil on an outcrop of sandstone on the northwest-facing slope of Peters Mountain at about 3,000 feet above sea level. It is growing in a deciduous forest with a few pines and appears to grow most vigorously in full sunlight, although it does tolerate shade. The soil is dark with a high organic content. LIFE HISTORY: Peters Mountain mallow reproduces by seeds and asexually by rhizomes. Flowering takes place from June through August and fruiting from July through October. The flowers are apparently pollinated primarily by sweat bees of the genus *Halictus*. Germination is enhanced by fire. THREATS: Construction of a now-rerouted section of the Appalachian Trail in the late

1960s and early 1970s played a major role in the decline of this species. The population is now carefully managed. The population is fenced to prevent browsing by white-tailed deer and watered artificially. Encroaching vegetation is periodically removed. NOTES: A number of plants are grown in an experimental garden at Virginia Tech. The Nature Conservancy owns a portion of the tract on Peters Mountain where this species occurs. Peters Mountain mallow is federally listed as endangered.

Abridged from "Peters Mountain Mallow" by Duncan M. Porter, pp. 130–133 in *Virginia's Endangered Species*, coordinated by Karen Terwilliger, 1991 (Blacksburg, VA: The McDonald & Woodward Publishing Company).

Ginseng

Panax quinquefolium

THREATENED

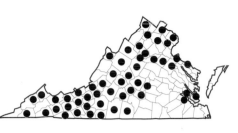

DESCRIPTION: Ginseng is a perennial herb arising from a large, spindle-shaped and often forked tuberous root. Stems are simple, 8 to 24 inches high. The three to five leaves, arranged as a whorl, are palmately compound. Leaflets are elliptical to obovate and toothed. Flowers 5-parted, white to greenish white; drupes red, occurring as clusters about 2½ inches in diameter. Ginseng may be confused with dwarf ginseng, *Panax trifolius,* and wild sarsaparilla, *Aralia nudicaulis.* The former is only 4 to 8 inches tall and has yellowish fruits. The latter stands 8 to 24 inches high, has twice-compound leaves, and purple to black fruits. DISTRIBUTION: Ginseng or sang was once widespread in eastern North America but is now uncommon as a result of overcollecting. In Virginia this species is known from most counties in the Appalachian Mountains, many on the Piedmont, and a few in the Coastal Plain. HABITAT: Ginseng is small herb of the floor of rich deciduous forests. THREATS: Exploitation of this plant for folk medicine has led to its rareness throughout its range. It has been exported from the New World to Hong Kong, China, and Korea since the eighteenth century. In the Far East it has long been used as a mild stimulant and aphrodisiac. Virginia exports between 7,000 and 10,000 pounds of wild-collected plants annually. NOTES: Control of the trade in ginseng is probably the best way to insure that its decline in abundance will be stopped. Buyers must now be licensed, and a harvest season has been set from August 15 to December 31. Collectors are encouraged to gather only plants that are in fruit and to plant the fruit in the same spot after collection of roots. Cultivation is quite easy and is encouraged as a way to relieve pressure on the wild population. Cultivated material is not subject to the Endangered Plant and Insect Species Act and may be harvested at any time.

Abridged from "Ginseng" by Duncan M. Porter, pp. 133–134 in *Virginia's Endangered Species*, coordinated by Karen Terwilliger, 1991 (Blacksburg, VA: The McDonald & Woodward Publishing Company).

Mat-Forming Water-Hyssop

Bacopa innominata

ENDANGERED

DESCRIPTION: Mat-forming water-hyssops are prostrate perennial herbs that form mats 2 to 12 inches in diameter. Stems are succulent, glabrous, and much-branched, rooting at nodes. Leaves are opposite, tiny, thick, round-ovate, and sessile. Flowers are axillary, corolla tubular, five-lobed, and whitish. Mat-forming water-hyssop is most apt to be confused with water-hyssop, *Bacopa rotundifolia*, which is found in the same general area. The latter has rounder leaves, $^4/_{10}$ to $^8/_{10}$ inch long. **DISTRIBUTION:** Mat-forming water-hyssop is known from along the Chickahominy, Mattaponi, and Pamunkey rivers. It has not been collected in Charles City County since 1941. Only seven localities are known in the Commonwealth. **HABITAT:** This plant grows along several tidal tributaries of the Chesapeake Bay on narrow shores or bordering mixed freshwater marsh, both subject to inundation by tides. These shores are in the upper meandering sections of the rivers. The substrate varies from soft and silty to a very fine sand-gravel-clay matrix. The plants occur with a number of widespread aquatic species, and also with some that are rare and of special concern, including Long's bittercress (*Cardamine longii*), Parker's pipewort (*Eriocaulon parkeri*), and yellow cowlily (*Nuphar sagittifolium*). **LIFE HISTORY:** Flowering is from late July to October. The plants are apparently self-pollinating. **THREATS:** Although mat-forming water-hyssop grows in a naturally disturbed habitat, unusual disturbance may destroy the plants or allow other species to outcompete them. Shoreline development, runoff, siltation, pollution, dredging, spoil banks, and disruption from boat waves are all possible threats. There is an evident decline in this species throughout its range. **NOTES:** Until recently, *Bacopa innominata* in Virginia and Maryland was known as *Bacopa stragula* and was considered to be endemic to several tributaries of Chesapeake Bay. It has been determined that the *Bacopa stragula* specimens are intertidal variants of *Bacopa innominata*, which also occurs in non-tidal freshwater habitats. There is a proposal to delist this species.

Abridged from "Mat-Forming Water Hyssop" by Duncan M. Porter, pp. 141–143 in *Virginia's Endangered Species*, coordinated by Karen Terwilliger, 1991 (Blacksburg, VA: The McDonald & Woodward Publishing Company).

Smooth Coneflower

Echinacea laevigata

ENDANGERED

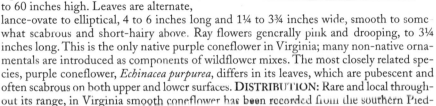

DESCRIPTION: Smooth coneflower has a single stem arising from a short taproot or vertical caudex. It stands 20 to 60 inches high. Leaves are alternate, lance-ovate to elliptical, 4 to 6 inches long and 1¼ to 3¾ inches wide, smooth to somewhat scabrous and short-hairy above. Ray flowers generally pink and drooping, to 3¼ inches long. This is the only native purple coneflower in Virginia; many non-native ornamentals are introduced as components of wildflower mixes. The most closely related species, purple coneflower, *Echinacea purpurea*, differs in its leaves, which are pubescent and often scabrous on both upper and lower surfaces. DISTRIBUTION: Rare and local throughout its range, in Virginia smooth coneflower has been recorded from the southern Piedmont and Ridge and Valley provinces. The total

number of recently verified sites is less than five, although an extensive survey of historic locations has not been conducted. HABITAT: Smooth coneflower grows in soils developed in limestone, diabase, and gabbro. It requires relatively open areas; natural disturbances, particularly fire, may have maintained significant areas of open habitat in the past. LIFE HISTORY: This perennial flowers in summer. Much remains to be learned of its biology. THREATS: Smooth coneflower is threatened by the loss of naturally-open habitat, which happens when natural disturbances are suppressed. Another threat is the collecting of individual plants. The roots of *Echinacea* species are highly valued for their pharmacological properties, and many members of the genus are collected for horticultural uses. NOTES: Even though the species utilizes roadside habitats and other areas disturbed by humans, its viability at these sites is unknown. Smooth coneflower was federally listed as endangered in 1989.

Abridged from "Smooth Coneflower" by J. Christopher Ludwig, pp. 144–145 in *Virginia's Endangered Species*, coordinated by Karen Terwilliger, 1991 (Blacksburg, VA: The McDonald & Woodward Publishing Company).

Virginia Sneezeweed

Helenium virginicum

DESCRIPTION: The Virginia sneeze-weed is a perennial herb with winged stems, 16 to 45 inches tall. Leaves are entire and without conspicuous lateral veins, the lower ones being nearly oblong and persistent. Inflorescences are several on sparsely-leafed peduncles. Ray flowers yellow, $4/10$ to $6/10$ inch long, and bearing three terminal teeth; disc flowers yellow. Virginia sneezeweed is similar to the sneezeweed, *Helenium autumnale*, but differs in having a sparsely-leafed stem and persistent basal leaves. **DISTRIBUTION:** The Virginia sneezeweed is endemic to Augusta and Rockingham counties, where it is known from 11 localities. **HABITAT:** Virginia sneezeweed inhabits wet meadows, sinkhole ponds, and wet depressions, habitats that experience dramatic changes in soil moisture, from inundation for eight months of the year to dry by early autumn. **LIFE HISTORY:** Virginia sneezeweed is a fibrous-rooted perennial that flowers from July through November. It is assumed to reproduce sexually. **THREATS:** The moist habitats are subject to drainage. ORV traffic may also be a threat. The populations at two sites are known to be extirpated. Cattle grazing is not a serious threat since the plant is not palatable to cattle, but trampling may be a threat. Grazing may, in fact, help control competing vegetation. **NOTES:** This species was listed as endangered in Virginia in 1989 and is a candidate for federal endangered or threatened status.

Abridged from "Virginia Sneezeweed" by Miles F. Johnson and Duncan M. Porter, pp. 145–146 in *Virginia's Endangered Species*, coordinated by Karen Terwilliger, 1991 (Blacksburg, VA: The McDonald & Woodward Publishing Company).

Sun-Facing Coneflower

Rudbeckia heliopsidis

ENDANGERED (proposed)

DESCRIPTION: Sun-facing coneflower is a perennial herb with basal offshoots and stout woody rhizomes. Stems are 24 to 48 inches high, almost smooth or sparsely hairy, freely branched above to form an open, corymb-like arrangement of flower heads. Leaves are simple, those at stem base with petioles at least three times as long as the ovate blades; stem leaves alternate, ovate, smooth or sparsely hairy, toothed or nearly smooth on margin. Central portion of flower head is purplish, rays yellow, 6/10 to 1 inch long. **DISTRIBUTION:** Sun-facing coneflower is known in Virginia only on the Coastal Plain from Prince George County, where it has been found at just one locality. **HABITAT:** The rare sun-facing coneflower grows along moist roadside ditches on a moist open, weedy roadside right-of-way at the headwaters of a small drainage. The population continues back from the roadway beneath a stand of loblolly pine and sweetgum in sphagnous swales. This species may require acidic soils. It does best where it receives relatively high levels of sunlight. **LIFE HISTORY:** Sun-facing coneflower blooms in mid-summer and fruits ripen in late summer to early fall. Vegetative reproduction apparently plays an important role. Its continued occurrence at the Prince George County site for over 50 years testifies to its ability to persist in a disturbed habitat where timbering has probably taken place at least twice since it was first discovered. **THREATS:** If viable seeds are indeed produced by this population, roadside mowing before ripening and dissemination of seeds is a primary threat. However, mowing in late fall might serve to control woody competitors. Scraping of roadside ditches has eliminated some individuals. Changes in timbering practices or in drainage patterns could be detrimental. The use of herbicides or introduction of alien species such as weeping lovegrass would also present threats. **NOTES:** This population was discovered in 1939 and reported again in 1949. There were no other reports for over 30 years; the site was accidentally rediscovered in the fall of 1979.

Abridged from "Sun-Facing Coneflower" by Donna M. E. Ware, pp. 147–148 in *Virginia's Endangered Species*, coordinated by Karen Terwilliger, 1991 (Blacksburg, VA: The McDonald & Woodward Publishing Company).

Arthropods

Arthropods are the most numerous of all living organisms. Insects alone outnumber species of all other animals and plants collectively by a factor of at least ten. The combination of protective exoskeletons, high rates of fertility, and short generation times (in most cases) appears to have bestowed an incredible advantage on arthropods, and they have successfully exploited every possible habitat.

The three most conspicuous common traits of arthropods are the sclerotized (often calcified) exoskeleton which is subdivided by elaborate segmentation and articulation to facilitate movement; the specialization of the anterior body pole into a head region, the appendages of which are modified into mouthparts; and the general adoption of internal fertilization.

Arthropods probably exert greater impacts, both positive and negative, on the welfare of humans than do members of any other group of animals. Far from being "lower" organisms, arthropods include many species which are as specialized evolutionarily as birds and mammals. Some insects, in addition, have evolved complex societies not so greatly different from our own, but typified by instinctive behavior rather than by the ability to use learned information.

Our ignorance of this immense group is by no means confined to the faunas of tropical regions. There has never been (and may never be!) an inventory of the arthropods native to Virginia, where a reasonably educated guess would place the likely number of species at between 20,000 and 30,000. Many of these are minute forms — aquatic microcrustceans and minuscule mites and insects of the soil and litter milieu — groups requiring highly specialized techniques to collect and highly trained specialists to identify. For many kinds of arthropods, there are no scientists qualified for doing the work.

Except for those species confined to precise habitats (such as caves) or those linked to specific host plants, the problems of adequate collecting almost preclude evaluation of the security of specific populations. Even the coverage of such organisms as butterflies and dragonflies falls far short of the treatment possible for virtually all vertebrates.

The great structural and biological diversity of arthropods, as well as the enormous number of species, have mandated that taxonomic groups be treated individually in this chapter and that each report on a taxonomic group be the work of a different specialist. For the most concise and complete summation of living arthropods in one source, the reader is referred to the *Synopsis and Classification of Living Organisms*, which is the standard for systematic arrangement used in this chapter.

Subphylum Chelicerata

Chelicerate arthropods are distinctive, fundamentally set apart from other arthropods by the organization of the head segments and associated appendages.

Chelicerates have no antennae. They are adapted primarily for capturing and holding food objects and lack the chewing ability enjoyed by tracheates and crustaceans. Of the three classes of chelicerates, two (Merostomata, horse-shoe crabs; and Pycnogonida, sea spiders) are exclusively marine and occur in Virginia only in the Chesapeake Bay. The third group, Arachnida, is predominantly terrestrial and is represented in Virginia by such organisms as spiders, daddy-long-legs, ticks, pseudoscorpians, and mites, in all perhaps about 2,000 species, many of which are conspicuous and well known to the nonscientist.

Traditionally, arachnids are said to have four pairs of legs (the textbook distinction from six-legged insects), but this is not strictly correct since the first true pair of legs are called "pedipalps" by arachnologists. The size and function of the pedipalps varies from group to group. In some arachnids, they are larger than the walking legs.

A common feeding strategy throughout the class Arachnida, virtually all of which are carnivores, entails holding the food object against the mouth and exposing it to an enzymatic saliva. Most digestion is therefore external (very rare among animals), and the resultant nutrient solution is drawn inside by a pump-like pharynx. Most arachnids are simply vagrant hunters; only spiders have exploited the opportunities of building nets and snares with sticky silk of their own manufacture. Not even all spiders are sedentary web-tenders; many, such as wolf spiders and jumping spiders, have abandoned that lifestyle to resume the ancestral arachnid technique of stalking and pouncing.

Most of the daddy-long-legs and pseudoscorpions likely to occur in Virginia have been found at least once. This cannot be said about spiders, of which we have records for only half of the 800 probable native species, nor about mites, upon which not even a beginning has been made.

In light of pervasive ignorance of arachnids in Virginia, these organisms cannot even be considered for threatened or endangered status. By far the greatest majority of species in Virginia are inhabitants of woodland environments (leaf litter, soil, low foliage) and are likely to be threatened only if extensive deforestation or urbanization takes place in a critical area, such as the region of Virginia Beach. Several spiders and pseudoscorpions are known solely from single caves in Virginia, but whether their total geographic range is so limited remains unclear. Many of these cave species may in fact be interstitial residents for whom a cave is simply a large crevice and would likely be unaffected by the negative impacts to cave ecosystems that would eliminate aquatic life.

Some arachnid species may occur in Virginia only in very small populations, at Virginia Beach, for example, or in the spruce-fir forest on Mount Rogers. The first-named site may well be one of concern for the survival of arachnids; the second would appear to be entirely secure under present conditions.

Subphylum Crustacea

Crustaceans are superficially distinguished by the presence of two pairs of antennae and a pair of appendages (mandibles) adapted for chewing. Many additional post-oral appendages may be variously modified as mouthparts. The appendages — at least those of generalized forms — are composed of two elements and are therefore described as biramous. Virtually all crustaceans are aquatic and depend on some kind of gill structure for gas exchange; the circulatory system consequently retains a role in oxygen transport.

There are about 33 generally accepted orders within Crustacea, almost all of which are marine. Many have evolved freshwater species, and at least several families of Isopoda have adapted to strictly terrestrial environments. Most crustaceans are less than half an inch long. Vast numbers of ostracods and copepods are microscopic in size. In all aquatic environments, small crustaceans are an important element in food chains. Most are herbivores or omnivores, some are obligate carnivores, and a few have become parasites.

Many of the orders of Crustacea occur in the freshwater biotopes of Virginia, but of these the smaller forms remain very poorly known. Copepods, cladocerans, and conchostracans surely are represented by hundreds of species still unrecorded. It is only the membership of the three large orders — Amphipoda, Isopoda, and Decapoda — which has been studied in Virginia. All are represented in freshwater streams and lakes; many amphipods and isopods also occur in groundwater, caves, and springs. The latter are of course most vulnerable to environmental changes because their ranges are small and localized and because their biotopes can be chemically poisoned or outright destroyed through urbanization and other forms of surface alteration. Although most of the large decapods of Virginia (crayfish, freshwater shrimp) tend to be more widely dispersed epigean forms, even they are at risk from environmental stresses.

Subphylum Tracheata

Tracheate arthropods share with crustaceans the characters of mandibular mouthparts but differ in having only one pair of antennae, uniramous appendages, and elaborate tracheal systems. Virtually all species are terrestrial; only a few have adapted secondarily to live in the intertidal zone and even fewer to life at the sea surface.

The subphylum Tracheata traditionally is viewed as containing the five classes Pauropoda, Symphyla, Chilopoda, Diplopoda, and Insecta. Pauropods and symphilids are tiny soil-inhabiting organisms that are rarely seen except by specialists. Some species are known only from single localities, while others are cosmopolitan. Because the state of knowledge of these small "myriapods" is so deficient, they must be passed over in this chapter.

Chilopods (centipedes) are generally larger and more familiar animals. All are predators and the only animals in which the first pair of legs (and not any appendages of the head) is modified into prominent poison fangs. So far some 3,000 species have been described; 52 are recorded from Virginia, but the data on distribution and population status are so deficient that this class is likewise not amenable to further consideration.

Diplopods (millipeds) are fairly large tracheates and not unfamiliar to gardeners and general naturalists. They are the only arthropods in which the embryonic somites have coalesced into pairs, giving the resultant impression of having two pairs of legs per segment in the adult. The class has been fairly well studied over the past several decades. Except for some forms considered to have very limited ranges, including some occupying cave systems, the majority of the 150 known in Virginia are not perceived to be in jeopardy.

Insects, in the broad sense, include tracheates with three pairs of thoracic legs, three well delimited body regions (head, thorax, abdomen), and, in most species, one or two pairs of wings. Despite the intensity with which insects have been studied for two centuries, there is still no consensus even about the number of orders. Current agreement, however, includes division of the winged forms into two subclasses: Paleoptera for mayflies and dragonflies, neither of which can fold the wings flat atop the abdomen; and Neoptera for all other orders, which do have the capability to fold the winds flat atop the abdomen.

Approximately 100,000 species of insects have been described or recorded from North America. Estimates for the entire world fauna range between one million and 50 million, the latter figure perhaps being the more likely. No inventory of insect species has been conducted for Virginia. One can postulate a total of 18,000 to 20,000 species of insects native to this state.

Knowledge of Virginia's insect fauna is extremely uneven. Depending on the existence of dedicated specialists and/or collectors, some groups (such as butterflies and dragonflies) are now reasonably well known. Other groups (Plecoptera, Heteroptera) have been studied to the extent that most native species have been found at least once, and the ranges of the more conspicuous can be mapped with some confidence. But for the great majority of insects, notably those of the four biggest orders (Coloptera, Diptera, Lepidoptera, and Hymenoptera), probably only half of the species in Virginia have been recorded so far and even these from one or two places only.

Those species having at least part of their life cycle in freshwater are likely to be most vulnerable to adverse impacts on their environments. What is bad for dragonfly nymphs will likely be bad for the mussels, fish, and turtles that share the same habitat. In this case, habitat protection with mandatory upstream control of water quality would seem to be far more effective than point-site measures to control water quality. For a few insects protection of food plants could be required, to the benefit of both insect and plant species.

Abridged from "Arthropods" by Richard L. Hoffman, pp. 173–180 in *Virginia's Endangered Species*, coordinated by Karen Terwilliger, 1991 (Blacksburg, VA: The McDonald & Woodward Publishing Company).

Madisons Cave Amphipod

Stygobromus stegerorum

THREATENED

DESCRIPTION: Blind, unpigmented cave-dwelling amphipod with long, slender antennae. Maximum body length approximately ¼ inch. Best distinguished by reference to original description and figures by Holsinger (1978) and the accompanying photograph. **DISTRIBUTION:** This species is known only from two caves in the eastern side of Cave Hill, near Grottoes, Augusta County. **HABITAT:** This amphipod inhabits two deep groundwater lakes in Madisons Cave and a similar lake in Stegers Fissure. These lakes are believed to be parts of a restricted aquifer existing beneath the southern end of Cave Hill. They provide a unique habitat which has not been observed in other caves of the area. One other crustacean, the threatened Madisons Cave isopod, also occurs in these lakes. **LIFE HISTORY:** Literally nothing is known about the life history of this extremely rare species. It is believed to feed on microorganisms and triturated organic matter. **THREATS:** The primary threat to this species is pollution of the aquifer beneath Cave Hill. Any disturbance of the sinkhole recharge area might well interfere with the passage of water and nutrients into the groundwater ecosystem. A threat will exist as long as much of the surface area of the south end of Cave Hill remains unprotected from human activities such as further construction, clearing of the woodland, or dumping of wastes into sinkholes. Disturbance of the lakes from inside the caves could also be detrimental. Fortunately the cave is being managed as a nature preserve with access permitted only for limited scientific study or educational purposes. **NOTES:** One sinkhole and a substantial piece of property above the cave have been altered to some extent with the construction of a water storage tank by the Town of Grottoes in 1982–1983. Fortunately, after long negotiations among Grottoes, the Virginia Cave Board, the US Fish and Wildlife Service's Office of Endangered Species, and the Farmers Home Administration, a plan to minimize adverse impacts on the recharge zone was developed. Uncontrolled access to Madisons Cave is prevented by a steel gate at the cave entrance installed in 1981 under the direction of the Cave Conservancy of the Virginias. (No photograph available.)

Abridged from "Madisons Cave Amphipod" by John R. Holsinger, pp. 181–182 in *Virginia's Endangered Species*, coordinated by Karen Terwilliger, 1991 (Blacksburg, VA: The McDonald & Woodward Publishing Company).

Madisons Cave Isopod

Antrolana lira

THREATENED

DESCRIPTION: An eyeless, unpigmented isopod; head with two pairs of antennae and mouthparts. Body length of adult females approximately ¹/₂ to ⁷/₈ inch; males ³/₈ to ⁵/₈ inch. Best distinguished by reference to original description and figures by Bowman (1964) and accompanying photograph. **DISTRIBUTION:** This species is known only from two caves in the eastern side of Cave Hill near Grottoes, Augusta County. **HABITAT:** This amphipod is a cave dweller inhabiting two deep groundwater lakes in Madisons Cave and a similar habitat in Stegers Fissure. These lakes are believed to be parts of a restricted groundwater deposit beneath the southern end of Cave Hill. In both caves, it coexists with the threatened Madisons Cave amphipod, a much less frequently seen crustacean. **LIFE HISTORY:** Primary food sources appear to be organically enriched silt and small particles of decaying wood. The population structure is skewed toward larger (?mature) animals, dominated by females; but very little is known about the reproduction or life history of the species. It is likely that individuals have an extended life span. **THREATS:** The extremely localized distribution of this species, coupled with its apparent restriction to a single, confined aquifer, make it highly vulnerable to disruptive human activities. Disturbance to the surface area above the caves through construction, clearing of woodland, or dumping of wastes into sinkholes could pollute the groundwater or interfere with the passage of water and nutrients to this unique groundwater ecosystem. Disturbance of the lakes from inside the cave could be detrimental also. **NOTES:** Because the Madisons Cave isopod is the only member of its genus and sole subterranean freshwater representative of the predominantly marine family Cirolanidae in America north of Texas, it is of great biogeographic interest. Its geographic isolation in the Valley of Virginia is curious inasmuch as all other subterranean cirolanid isopods live in areas that either are near coastal marine zones at present or were exposed to shallow marine embayments during the Cretaceous or Tertiary periods.

Abridged from "Madison Cave Isopod" by John R. Holsinger, pp. 186–188 in *Virginia's Endangered Species*, coordinated by Karen Terwilliger, 1991 (Blacksburg, VA: The McDonald & Woodward Publishing Company).

Lee County Cave Isopod

Lirceus usdagalun

ENDANGERED

DESCRIPTION: The Lee County cave isopod is an eyeless, unpigmented species measuring $^2/_{10}$ to $^3/_{10}$ inch in length. The body is about 64 percent longer than wide, and the head is about $^1/_3$ as long as wide with deep incisions on its lateral margins. **DISTRIBUTION:** The Lee County cave isopod is endemic to Virginia, restricted to four caves in south-central Lee County. Three of the four caves are hydrologically integrated into a single system; the fourth is located about six miles northeast of that system. However, the Lee County cave isopod has been extirpated from this last location as a result of groundwater pollution. **HABITAT:** *Lirceus usdagalun* inhabits caves developed in a band of low-dipping, middle Ordovician limestone on the southern flank of the Cedar Syncline. It is associated with the gravel or rock substrate of streams and seep pools in the caves. **LIFE HISTORY:** No information is available on the life history of this species. **THREATS:** The Lee County cave isopod is particularly susceptible to contamination of groundwater from surface pollutants passing directly through sinkholes and fissures in the limestone terrain. This species has already been extirpated from one of the two cave systems it originally occupied. **NOTES:** The Lee County cave isopod is a troglobite or obligate cave-dweller. Like many other species restricted to cave environments, it is distinguished by the loss of eyes and pigment and the attenuation of the body. Species in the genus *Lirceus* occur in the eastern and mid-western United States and the Great Lakes region of southern Ontario, Canada, in a variety of habitats including springs, seeps, streams, ponds, sloughs, and drain outlets. While some of these other species have been found in cave streams, only *Lirceus usdagalun* from Lee County and a closely related species in neighboring Soctt County, Virginia, lack eyes and pigment and are considered obligate cave-dwellers.

Prepared by Rick Reynolds, Virginia Department of Game and Inland Fisheries, and John R. Holsinger, Old Dominion University, March 1994.

Ellett Valley Pseudotremid Milliped

Pseudotremia cavernarum

THREATENED

DESCRIPTION: A medium-sized milliped with maximum body length of about 1 inch. Slender body with long slender legs; exoskeleton largely unpigmented and pale in color. DISTRIBUTION: This milliped is known only from caves in Ellett Valley, east of Blacksburg, Montgomery County. HABITAT: This species occupies small caves and associated cracks, crevices and fissures. LIFE HISTORY: Essentially nothing is known about the biology of this milliped. Specimens were rarely seen at the type locality except during April, when literally hundreds appeared on the walls and floor of the cave and mating was observed. Presumably at other times, the millipeds retreat into small crevices. Loss of pigment and other characteristics confirm this species as an obligate cave-dweller. THREATS: The major jeopardy to this species is surface development, such as filling of sinkholes, which would interfere with or interrupt movement of organic nutrients into cave environments. NOTES: The type locality, Erharts Cave at Ellett, Montgomery County, was developed during the last century as an attraction for guests at nearby Yellow Sulphur Springs. The original specimens were obtained by E. D. Cope, who sojourned with his family in southwest Virginia in the 1860s. In the late 1950s a limestone quarry was opened nearby, and within a decade the entire hill and the cave it contained were reduced to gravel. (No photograph available.)

Abridged from "Ellett Valley Pseudotremid Milliped" by Richard L. Hoffman, pp. 190–191 in *Virginia's Endangered Species*, coordinated by Karen Terwilliger, 1991 (Blacksburg, VA: The McDonald & Woodward Publishing Company).

Laurel Creek Xystodesmid Milliped

Sigmoria whiteheadi

THREATENED

DESCRIPTION: The surface of this milliped is smooth and polished, nearly black with bright lemon yellow spots connected by a thin transverse yellow band along the back edge of each segment. Legs and antennae are a less intense yellow, tinged with beige. **DISTRIBUTION:** This species is known only from the type locality near the headwaters of Laurel Creek on the Blue Ridge Parkway, Patrick County. **HABI-TAT:** This milliped was found under mixed rhododendron and hardwood leaf litter on sandy substrate about 10 to 15 feet from a stream. Red maple with some oak, hemlock, and dogwood comprise the canopy. **LIFE HISTORY:** Virtually nothing is known about the biology of this milliped. **THREATS:** At present, since the type locality is inside the property boundary of the Blue Ridge Parkway, no overt threat is foreseen. The apparently restricted range and small population size suggests that this milliped may be disappearing, perhaps reflecting the natural extinction of a relict species. **NOTES:** Millipeds of the family Xystodesmidae frequently occupy very limited ranges, and many species are known from only single localities (often because of inadequate collection). In the case of this species, a number of apparently suitable sites have been examined, both near the type locality and farther to the southwest, without success. The closest relative of the Laurel Creek species seems to be *Sigmoria austromontis*, which is endemic to the South Mountains of North Carolina.

Abridged from "Laurel Creek Xystodesmid Milliped" by Richard L. Hoffman, pp. 191–192 in *Virginia's Endangered Species*, coordinated by Karen Terwilliger, 1991 (Blacksburg, VA: The McDonald & Woodward Publishing Company).

Cherokee Clubtail

Stenogomphurus consanguis

THREATENED (proposed)

DESCRIPTION: A small dragonfly with body length of about 1¾ to 2 inches, hindwing length of 1⅛ to 1⅜ inches. Hindwings have one postmedian crossvein. Green with wings tinged with brown. DISTRIBUTION: The only two known localities in Virginia are a small tributary of Spring Creek in Washington County and Copper Creek in Scott County. HABITAT: This species seems to prefer small permanent streams which often drain small ponds. Nymphs are shallow burrowers in the small accumulations of silt along pool areas. Adults are seldom seen far from streamside. LIFE HISTORY: Flight season is from 20 May to 15 June. The post emergence feeding period is apparently short, lasting about one week, at which time adults are found near the parent stream. Territorial behavior of the males primarily involves perching on low vegetation near likely oviposition sites. Females oviposit near the banks of pool areas, where they tap the tip of the abdomen to the water surface between brief periods of hovering. The nymphs are shallow burrowers in soft bottom mud; the nymphal stage probably lasts two years.

THREATS: The nymphal stages of the Cherokee clubtail are vulnerable to adverse alterations of water quality, such as siltation, thermal changes, chemical pollution, and deoxygenation, as well as impoundments and channelization. NOTES: The Cherokee clubtail's distribution is apparently centered about the upper Tennessee River Valley. It is likely that many populations have been lost due to impoundments throughout its range.

Abridged from "Cherokee Clubtail" by Frank Louis Carle, pp. 208–209 in *Virginia's Endangered Species*, coordinated by Karen Terwilliger, 1991 (Blacksburg, VA: The McDonald & Woodward Publishing Company).

Nantahala Belted Skimmer

Macromia margarita

THREATENED (proposed)

DESCRIPTION: A medium-sized drag-onfly with body length of about 3 inches and hindwing length of about 2 inches. Compound eyes contiguous dorsally; forewing triangle usually two-celled. DISTRIBUTION: The only known Virginia popula-tion occurs in the Little River in Floyd County. HABITAT: In small and medium sized rivers of the southern Blue Ridge physiographic province, nymphs crawl through roots, rocks, and detrital accumulations in pool areas; well aerated water seems to be a require-ment. Adult feeding areas are apparently in open glades near the river. LIFE HISTORY: Flight has been observed from 3 June to 15 June. The post emergence feeding period lasts about three days. Territorial behavior of the males is typified by continuous rapid flight up and down river about two feet above the water; the territory may extend for more than 300 feet. Oviposition usually occurs at the head of a riffle area. Females fly in rapidly and follow a zigzag course while tapping the water with the abdomen. After about five such taps they will dart back to the shore and rise to tree top level to hang up and rest. Nymphs are sprawlers and climb spiderlike through accumulations of debris caught up along the river. The nymphal stage probably lasts two years. THREATS: Nymphs are vulnerable to any disturbance that results in poor water quality — siltation, thermal fluc-tuation, deoxygenation, and chemical pollution. Of these factors, siltation is the most pervasive threat in the Little River. NOTES: This dragon-fly has the most restricted range of any North American member of its family. It is known only from the Culla-saja, Little, and French Broad rivers in North Carolina, from streams in Lumpkin County, Georgia, and Pickens County, South Carolina, and the Little River in Virginia. The Vir-ginia population is by far the most distant from the center of the range.

Abridged from "Nantahala Belted Skimmer" by Frank Louis Carle, pp. 209–210 in *Virginia's Endangered Species*, coordinated by Karen Terwilliger, 1991 (Blacksburg, VA: The McDonald & Woodward Publishing Company).

Swamp Skimmer

Tetragoneuria spinosa

THREATENED (proposed)

DESCRIPTION: A small dragonfly with body length of about 1⅞ to 2 inches and hindwing length of about 1 to 1½ inches. Eyes contiguous dorsally. Forewing triangular, interspace constricted toward wing margin. **DISTRIBUTION:** Two localities are known in Virginia: one in Great Dismal Swamp National Wildlife Refuge; another bordering the Nottoway River in Southampton County. **HABITAT:** Open swamps with movement of water are preferred. The nymphs apparently crawl through accumulations of detritus along the deeper waters of swamplands. Adult feeding areas are along dirt or sand roads or in nearby open glades. **LIFE HISTORY:** The two collections of adults in Virginia were made in mid-April. The post emergence feeding period probably lasts about three days. Territorial behavior of the males occurs along the edges of the swampland; flight is about three feet above the water. During oviposition, which typically occurs in these same areas, the females tap the water once or twice, leaving a long string of eggs to uncoil. The nymphs are sprawlers on mud and detritus; the nymphal period is probably two years. **THREATS:** The swamp skimmer's swampland habitats are particularly threatened by channelization, draining, and lumbering. **NOTES:** The swamp skipper is known from New Jersey, Delaware, Maryland, Virginia, North Carolina, South Carolina, Georgia and Alabama, but rarely is it abundant anywhere. The dragonfly's early flight season may account in part for the paucity of locality records.

Abridged from "Swamp Skimmer" by Frank Louis Carle, pp. 212–213 in *Virginia's Endangered Species*, coordinated by Karen Terwilliger, 1991 (Blacksburg, VA: The McDonald & Woodward Publishing Company).

Northeastern Beach Tiger Beetle

Cicindela dorsalis dorsalis

THREATENED

DESCRIPTION: Body length is about ⁵/₈ inch. Wing covers are either completely white or, more frequently, with several fine grayish green lines on each. Ventral surfaces of thorax and abdomen grayish green, with dense white hairs on sides. DISTRIBUTION: Populations are known from 25 areas on both the eastern and western shores of the Chesapeake Bay. HABITAT: Preferred habitat is wide, white sandy beaches along the shore of the Chesapeake Bay where there is low to moderate human use or disturbance. Dynamic point or exposed beaches where there is active build up of sand appear to be favored. LIFE HISTORY: Adult beetles of this species are active from early June to early September along the wet intertidal area of beach where they forage for small insects and amphipods or scavenge on crustacean or fish carrion. The adults are active during summer days unless cloudy and cool conditions prevail. Eggs are laid in the upper intertidal zone, where the larvae live in well developed burrows, waiting at the entrances to seize prey which passes nearby. A two year life cycle requires that larvae overwinter twice in the beach habitat. THREATS: The dramatic decline of this species in the northeastern US is probably the result of habitat destruction and disturbance by human activities. Beetle activity is disrupted by heavy use of beaches by foot or vehicular traffic. A survey on Assateague Island suggested that the impact of offroad vehicles is particularly

damaging. Coupled with natural limiting factors such as winter storms, beach erosion, and natural predators, such human impacts could cause major declines in larval recruitment and the subsequent extinction of local populations. NOTES: The northeastern beach tiger beetle historically ranged from the Chesapeake Bay north to Massachusetts, but presently its distribution is limited to Virginia and Maryland and one population on Martha's Vineyard, Massachusetts. Beach nourishment and shoreline stabilization, proposed for several sites, might be destructive to larvae and may render the habitat unsuitable for subsequent larval recruitment and development. The species seems very susceptible to local population extinction. Preservation measures will require protection of a series of adjacent or nearby sites in any given area. Three favorable areas that should be targeted for protection are Smith Point to the Kilmarnock area, Gwynns Island to New Point, and Savage Neck to Picketts Harbor.

Abridged from "Northeastern Beach Tiger Beetle" by C. Barry Knisely, pp. 233–234 in *Virginia's Endangered Species*, coordinated by Karen Terwilliger, 1991 (Blacksburg, VA: The McDonald & Woodward Publishing Company).

Rare Skipper
Problema bulenta

THREATENED (proposed)

DESCRIPTION: The rare skipper has a maximum wing span of about 2½ inches. Both sexes are yellow to golden dorsally with wide dark margins on both forewings and hindwings; ventrally wings are golden yellow, hindwings lacking a dark margin. The broad brown margins of upper wings and immaculate underwings are diagnostic. Caterpillar is undescribed. DISTRIBUTION: The single known Virginia population of this skipper inhabits a marsh along the Chickahominy River south of Lanexa in New Kent County. HABITAT: This skipper is known only from slightly brackish river marshes. LIFE HISTORY: The rare skipper's life history is poorly known. Host plant to the caterpillar is big cordgrass, *Spartina cynosuroides*. Saltmarsh cordgrass, *Spartina alterniflora*, and common reed, *Phragmites australis,* may also be food plants. Adults are known to take nectar from swamp milkweed (*Asclepias incarnata*), pickerelweed (*Pontederia cordata*), and buttonbush (*Cephalanthus occidentalis*). THREATS: The marsh habitat of the rare skipper is threatened by development for housing and recreation which may adversely alter the hydrology or erosional patterns. Increased recreational use of boats on the Chickahominy River is resulting in boat-wake damage to the river banks and marshes in some areas. NOTES: The rare skipper is known from widely separated areas along the Atlantic Coast from New Jersey to Georgia. It was first described in 1834 and then not collected again for nearly a century. The Virginia colony was discovered by J. Bauer and B. Dixon in 1967. Populations were discovered in New Jersey in 1989.

Abridged from "Rare Skipper" by Christopher A. Pague, pp. 238–240 in *Virginia's Endangered Species,* coordinated by Karen Terwilliger, 1991 (Blacksburg, VA: The McDonald & Woodward Publishing Company), and updated with information from W. J. Cromartie and D. F. Schweitzer (1993) "Biology of the Rare Skipper *Problema bulenta* (Hesperidae) in southern New Jersey," *Journal of the Lepidopterists' Society* 47(2): 125-133.

Regal Fritillary

Speyeria idalia

THREATENED (proposed)

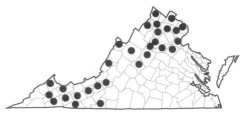

DESCRIPTION: Forewings dorsally yellowish orange with narrow black bars and a single row of small black dots in both sexes. Forewings of male have black edge and marginal row of orange spots; forewings of female are black with six to ten white spots and no orange spots along rear margin. Hindwings dorsally black in both sexes but paler in males, in which there is a central row of white spots and a submarginal series of five orange dots. In females the submarginal row of dots is white instead of orange. Underside of hindwing is black with conspicuous large white markings. The species is illustrated in all recent butterfly books. DISTRIBUTION: In Virginia this butterfly is restricted to the mountains and inner Piedmont. HABITAT: The regal fritillary is a grassland species. Open areas such as pastures which contain wet patches or streams are preferred. Adequate nectar sources for the entire flight season are critical. Many species of flowers in many families are used for nectar; some favorites include thistles, milkweeds, red clover, and blazing star. LIFE HISTORY: Eggs are laid near violets; larvae hatch in late summer or fall and hibernate without feeding, probably among sheaths of fallen grass stems and leaf litter. After exposure to warm temperatures in spring, larvae emerge from hibernation and begin to feed; several species of violet serve as food plants. Larvae begin to mature in late May, and males appear about mid-June, females about two weeks later. Females typically have a longer larval period than males and generally persist well into August or even later. THREATS: The major causes of the decline of this butterfly species are very poorly understood. In other parts of the range, extirpation has been attributed to outright habitat destruction, to succession, to too much or too little fire, and to spraying to control gypsy moths. Even in areas when extirpation has occurred, seemingly suitable habitats often remain unoccupied. Perhaps habitats are becoming so widely scattered that the rate of local colonization cannot keep up with the rate of local extinction. Over-collecting is now considered a threat also. NOTES: The regal fritillary has declined over almost half of its previously vast range and is extirpated in several states and Canadian provinces.

Abridged from "Regal Fritillary" by Dale F. Schweitzer, pp. 243–245 in *Virginia's Endangered Species*, coordinated by Karen Terwilliger, 1991 (Blacksburg, VA: The McDonald & Woodward Publishing Company).

Mollusks

Freshwater Mussels

The freshwater mussel fauna in Virginia is among the most diverse in the United States and comprises about 73 species. Many are elements of a Cumberlandian fauna associated with the unglaciated areas of the Appalachian Plateaus Province. The great geologic age and lengthy isolation of the Tennessee and Cumberland rivers and their faunas promoted speciation within ancestral lineages, such that rivers with Cumberlandian species in combination with species originating from the Mississippi Basin contain freshwater faunas that are among the most diverse in the world. However, the abundance of some species has declined, while others have been extirpated. Declines in mussel diversity, distribution, and population levels are widespread throughout the eastern United States due to causes almost exclusively attributed to human activities. Most of Virginia's vulnerable mussel species reside in tributaries of the Tennessee River drainage system in the southwestern part of the state.

Freshwater mussels are vulnerable to the disturbance of instream habitat (caused by activities such as channelization, dredging, and sedimentation), riparian habitat (resulting from buffer strip cutting, bank destabilization, erosion, and so forth), and water quality (resulting from waste discharges, toxic spills, acidification, nonpoint pollution, and so on). Once a population of mussels is lost from a river or river reach, decades are required for recolonization to occur because mussels are relatively immobile and have unusual reproductive cycles.

Most species of mussels that occur in Virginia have similar life cycles and habitat requirements. During the spawning period, males release sperm that enter females through the incurrent aperture. Eggs are fertilized and embryos are retained in the female's gills until they develop into mature parasitic larvae called glochidia. Long-term brooders spawn in late summer, retain glochidia over the winter, and release them into the water the following spring and early summer. Short-term brooders spawn in spring and release glochidia in the summer.

Glochidia are obligate parasites and must attach to a suitable fish host. Some attach to the fins or epidermis of fish, others attach to gill filaments. The glochidia become encysted and metamorphose in one to three weeks into juveniles, which drop from the host to begin the dominant free-living phase of the life cycle. The natural history of the first three to four years of a cohort is essentially unknown; few juveniles have been collected in field samples. Mussels achieve sexual maturity in three to five years; some riverine species may live for more than 50 years.

It appears that relatively low but continuous recruitment sustains existing populations. The glochidia stage is interpreted as an adaptation for dispersal. Since the incidence and prevalence of infestations on fishes are low, rates of colonization or recolonization are expected to be particularly slow for this invertebrate group.

Most riverine species of mussel reside in runs and riffles of moderate-sized (third order) streams to large rivers. A stable substrate of mixed cobble, gravel, sand and silt with suitable water velocity and water quality provides optimal habitat for them.

Adult mussels are filter-feeders and remove silt, one-celled algae, zooplankton, and detritus from the water. A mussel bed functions, therefore, as a biological filter by removing organic and inorganic particulate matter from the water and improving water quality downstream.

The identification of freshwater mussel species requires primarily the use of shell characters (Figure 4). The shape and dimensions of the shell, coloration of epidermis and nacre, and sculpture of the beak and surface are among the more important diagnostic features. Many species, however, are polymorphic in shell characters such that differences among populations, indeed among individuals within the same population, can cause difficulties for even professional taxonomists. Confidence in species identification can only be achieved by experience or by submitting specimens to reputable malacologists for confirmation.

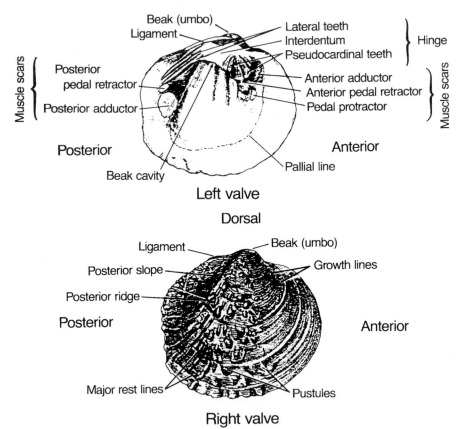

Figure 3. A generalized mussel shell showing selected morphological characters.

Snails

Three subclasses of gastropods occur in Virginia, but only terrestrial and fresh-water snails (Pulmonata and Prosobranchia) are addressed in this chapter. The former is distinguished by a mantle cavity modified to function as a lung and an aperture that is nonoperculate. The latter is characterized by a gilled mantle cavity and an operculum attached dorsally on the posterior of the foot that may be drawn up to seal the aperture of the shell.

Land Snails

Terrestrial habitats of snails range from moist wooded hillsides to dry rocky outcrops and from sandy vegetated beaches to subalpine zones on ridge tops. Most land snails overwinter under sheltered cover or buried in the ground. Climate at ground level, particularly temperature and humidity, is critical for land snails.

Some snail species are carnivores and feed on small soil arthropods and other species of snails. Most, however, are herbivores that consume plant and leaf material, or detrivores that consume decaying leaf litter. These species occur on or in the detritus and soil layers and can be collected by sifting this material.

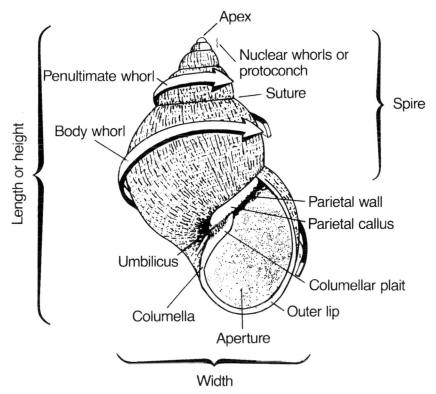

Figure 4. A generalized snail shell showing selected morphological characters.

67

The land snail fauna is extremely diverse in Virginia, where it is represented by 15 families, 39 genera, and about 187 species. Collection records for most counties are incomplete, and our knowledge of species distributions is somewhat fragmentary. Nevertheless, we do know that Virginia, because of geographic location and physiographic diversity, has one of the most diverse snail assemblages in the eastern United States. Many northern species reach the southern limit of their distributions in Virginia, while some southern species have their northern limits here. In addition, the mollusk fauna from the central United States extends eastward into southwestern Virginia. Each of Virginia's five physiographic provinces contains a relatively distinct fauna. The Cumberland Plateau in southwestern Virginia harbors more than 50 percent of the recorded snail species in Virginia. The Ridge and Valley has the next highest number of recorded species, followed by the Blue Ridge, Piedmont, and Coastal Plain provinces, respectively.

Of the roughly 187 species of land snails recorded in Virginia, 51 are known to occur in only one or two counties. Most of these species, however, also occur in adjoining states, and what appear to be limited distributions in Virginia are likely artifacts of inadequate sampling. Five species are endemic to Virginia. One of these, the Virginia fringed mountain snail, is federally listed as endangered.

Land snails are classified mainly by differences in internal body parts, particularly in the reproductive and excretory systems. The external shell is often useful in species identification (Figure 5), but shell morphology is so variable that field experience is required for positive identification.

Freshwater Snails

Most aquatic snails are vegetarian and feed primarily on the algae, diatoms, and desmids adhering to various substrata. Dead plant material and, occasionally, dead animal matter is ingested by some species.

Certain ecological factors have a pronounced effect on snail occurrence. In general, soft water contains few species whereas hard water has a greater diversity and abundance because the content of calcium carbonate essential for shell construction appears to be important for many species. In aquatic mollusks, pH of the water determines the availability of carbon dioxide to form calcium carbonate in shells and influences the capacity of the animal to deposit calcium. Most species require moderate to high levels of dissolved oxygen; hence polluted rivers and deep ponds and reservoirs usually have few, if any, snails. Typical habitats for snails are shallow waters (less than 10 feet deep) where sufficient food and a suitable substratum occur. Many species are restricted to certain types of substrata such as boulders, mud, or vegetation; others are more general in their requirements. Activity is greatest in spring and summer when feeding, growth, and reproduction are at their maxima. In winter, some species hibernate or remain inactive in pond and river bottoms whereas others remain active, but at reduced levels.

The only aquatic snail listed at this time is the Unthanks Cave snail. This endangered snail represents a new genus and species known only from one site, Unthanks Cave, in Lee County.

Differentiation of freshwater species is based typically on anatomical details of shell, radula, and reproductive organs. Except for the limpets, all aquatic snails have a spiral or discoidal coiled shell, usually drab in color, and many have fine growth lines and spiral sculpture. Most species are dextral. Variations in size, shape, thickness, and surface markings of the shell occur within genera and species and are attributed to environmental constraints. These phenotypic variants are particularly common in river snails.

Threats to Mollusks

Aquatic Species

Dams for industrial and municipal water and hydropower converted river reaches to large lakes, altered natural fish communities, increased sedimentation, and lowered oxygen levels and water temperatures. The increase in sedimentation is thought to have the most significant adverse impact on mussels indigenous to dammed streams.

Siltation resulting from poor land use and deforestation has altered riverine habitats and interferes with the respiration and feeding of aquatic organisms. Because mussels are sedentary, they are unable to move from habitats that are being gradually buried in silt. Siltation also reduces light penetration and adsorbs toxic substances for deposition, resulting both in a decline in oxygen concentration and in exposure of mollusks to toxicants. A combination of inorganic silt, coal waste deposits, and water quality degradation from mining activities in southwestern Virginia has eliminated most of the molluscan fauna from the upper Powell River and may jeopardize the remaining fauna in lower Lee County.

Water pollution by organic and inorganic contaminants is another major factor in the decline of species throughout the state. Chlorinated discharges from sewage treatment plants and electric generating plants have eliminated mollusks downstream of their effluents. Toxic spills in the Clinch River at Carbo in 1967 and 1970 eliminated most of the aquatic fauna for about 12 miles downstream. The mussel fauna has not recovered. Prolonged contamination of rivers by mercury (North Fork of Holston River, South Fork of Shenandoah River), mirex (James River), and other toxic chemicals (Elizabeth River) have also contributed to the loss of specific populations and impeded their recovery. The effects of pollutants on mollusks are poorly documented, but the reduction or elimination of mollusks downstream of industrial and municipal outfalls and other point source discharges probably results from poor water quality. Overcollecting by zealous shell enthusiasts, professional and amateur alike, has contributed to the decline of populations at well known locations such as Speers Ferry and Kyles Ford on the Clinch River.

The Asian clam, *Corbicula fluminea*, which now resides in every major river in the state, dominates the molluscan fauna in many rivers. It is not known whether native species will be able to coexist at population levels comparable to those that existed prior to invasion by this exotic species. The impending invasion of the zebra mussel, *Dreissena polymorpha*, is a serious threat to all freshwater mussels.

Terrestrial Species

Habitat disruption by changes in land use can severely affect local snail populations. Deforestation removes not only the protective cover of the trees but also effects major changes in microhabitats at the ground level where snails live. As trees are removed, greater insolation occurs, and the soil is heated. This results in greater evaporation and drying. In addition, deforestation reduces the amount of leaf litter, thereby limiting food resources. Disturbance of forest habitats can easily lead to local extirpation.

Another major disruption of local populations of snails is caused by widespread herbicide spraying. Herbicides tend to reduce the amount of available food resources and act as toxicants to the snails themselves. Perhaps of significant long-term concern is the acidification of habitats by acid precipitation. Snails require a steady source of calcium carbonate, most of which goes directly into shell production. As pH decreases, shell erosion increases; hence, more energy is required to sustain shell growth and maintenance.

Recommendations

Except for freshwater mussel surveys conducted in southwestern Virginia to determine the status of federally listed species, there has been no systematic survey of the other molluscan taxa in the ecosystems of Virginia. The paucity of recent mollusk information in Virginia is acute, and a survey of freshwater mussels in eastern Virginia and the snail fauna statewide is essential in order that rare species and concomitant habitats can be identified and protected.

Biological and ecological research on mollusks in Virginia has not been pursued sufficiently. Very few studies have addressed the life histories and ecological requirements of mollusks in Virginia. Regrettably, many species likely will be extirpated or extinct before the scientific community is able to assess their role in Virginia's ecosystems. It is imperative that both basic and applied research be conducted to provide the knowledge required to conserve and restore populations throughout the state.

The success of management and protection efforts is dependent on habitat protection and restoration. Most decreases in range and declines in abundance of species have been directly attributed to habitat loss and degradation. State agencies with the responsibility and authority to manage Virginia's land and water resources are well aware of the environmental threats to most species. However, the site-specific threats to terrestrial and aquatic mollusks need to be identified throughout the state and eliminated so that natural and human aided recovery can proceed. Because a high percentage of freshwater mussels is considered to be endangered or threatened, this faunal group is in immediate need of attention. The rivers of Virginia, particularly those in southwestern Virginia, should be closely monitored to sustain and improve water quality. A combination of habitat protection and information derived from research on the biology and ecology of these species is needed so that resource managers can assist with the conservation and recovery of depleted species.

Abridged from "Mollusks" by Richard J. Neves, pp. 251–263 in *Virginia's Endangered Species*, coordinated by Karen Terwilliger, 1991 (Blacksburg, VA: The McDonald & Woodward Publishing Company).

Spectaclecase

Cumberlandia monodonta

ENDANGERED

DESCRIPTION: The spectaclecase is characterized by an elliptical and elongate shell, three times as long as wide. It is unsculptured. Dorsal and ventral margins are nearly straight and parallel. Epidermis is dark brown to black, and the nacre is white to bluish white and iridescent posteriorly. Maximum length is about 7 inches. DISTRIBUTION: In Virginia, the spectaclecase is found only at two sites in the Clinch River in Scott County. It is now extremely rare at the Speers Ferry site. HABITAT: The spectaclecase occurs in clean, free-flowing rivers, where it is often embedded at the bases of large boulders or along cracks in ledges. It prefers a stable bottom of large cobbles or boulders and may be found at the river margin where the swift current of the mainstream meets quiet water at the edge of pools. This species seldom if ever moves except to burrow deeper. Generally it will die in place if exposed to low water conditions. LIFE HISTORY: The life cycle is presumably typical of freshwater mussels. There may be two broods in a season. No fish hosts are known. THREATS: The Clinch River has been affected several times by spills of toxic substances. Poor survival of recent mussel transplants in the Clinch River suggests that water quality problems may still be occurring. Alteration and destruction of stream habitat by impoundment, dredging, or channelizing are major threats to mussels in the Clinch River. Overzealous collecting of this unique species may also have contributed to its decline in Virginia. The impending invasion of the zebra mussel is a serious threat.

NOTES: The spectaclecase is a primitive species in a monotypic genus. It has a disjunct distribution in the central US, where it is known from the Cumberland, Tennessee, and Mississippi rivers and from rivers in Illinois, Indiana, Kentucky and Missouri. In recent years in Virginia it has declined in numbers to the degree that population sizes may have fallen below viable levels.

Abridged from "Spectaclecase" by Lisie Kitchel, pp. 264–266 in *Virginia's Endangered Species,* coordinated by Karen Terwilliger, 1991 (Blacksburg, VA: The McDonald & Woodward Publishing Company).

Dwarf Wedgemussel

Alasmidonta heterodon

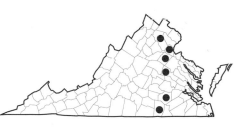

ENDANGERED

DESCRIPTION: The smallest freshwater mussel in Virginia's Atlantic drainages, the dwarf wedgemussel has a shell length rarely exceeding 1½ inches. Valves are subtrapezoidal in shape, with a distinct but broadly rounded posterior ridge. When well developed, the angle between the gently descending ventral margin and sharply descending posterior margin gives the shell its characteristic "wedge" shape. The smooth epidermis of the adult is brown to sometimes greenish olive, with green or tan rays visible in younger and pale-colored specimens. Valves are thin but relatively strong. Moderate sexual dimorphism is evident as the female is more inflated than the male in the posterior ridge region. Nacre is usually silvery white to bluish, often with cream, tan, or olive colorings toward the shallow beak cavity. This is the only North American freshwater mussel typically with two lateral teeth in the right valve and one in the left, a reversal of the normal pattern for freshwater mussels. DISTRIBUTION: The dwarf wedgemussel is currently known in Virginia from the Nottoway River, Aquia Creek, Cedar Run in the Occoquan Creek drainage, and the Po and South Anna rivers of the York drainage system. HABITAT: This mussel appears to prefer stable sand and gravel areas of streams, but may be found in a wide variety of stream environments. It is always associated with flowing water, even if the flow is mostly seasonal. It does not occur in strictly mud-bottomed pools. LIFE HISTORY: The dwarf wedgemussel is a long-term brooder. Gravid females occur from late August to early June; glochidia are released in the spring. Fishes native to its range and identified as hosts are the johnny darter, *Etheostoma nigrum,* and the tessellated darter, *Etheostoma olmstedi.* Glochidia may use sculpins, *Cottus* sp., as hosts also. THREATS: Loss of habitat and poor water quality are the main threats. This species does not occur in reservoirs. Siltation from land disturbance also is a problem. Competition from non-native bivalves may be an additional problem for this species. NOTES: The curiously reversed hinge teeth and small size make this species a very distinctive mussel. The dwarf wedgemussel was considered extirpated from Virginia until rediscovered in 1990.

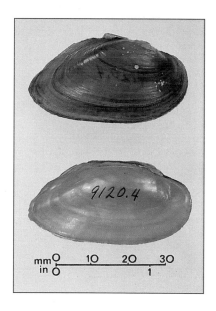

mm 0 10 20 30
in 0 1

Prepared by Sue A. Bruenderman, Missouri Department of Conservation, and Philip H. Stephenson, Richmond, VA, May 1994.

Brook Floater

Alasmidonta varicosa

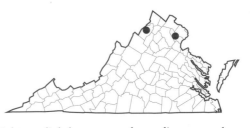

DESCRIPTION: A small mussel, the brook floater is generally 2 to 3 inches long. The shell is somewhat kidney-shaped in outline. The anterior margin is abruptly curved; ventral margin straight to slightly concave, descending somewhat. Posterior margin is biangulate below and broadly curved to straight above. The dorsal margin is slightly convex. Beaks are low and located well forward, approximately one fourth of length from anterior margin. Numerous short, low corrugations or ridges occur on posterior slope of shell. These ridges tend to be oriented radially. No other shell within its range has distinctive radial ridging on posterior slope. Shell is firm but not thick. Epidermis tends to be smooth centrally and rough elsewhere. In adults, epidermis is brownish with dark fine rays sometimes present; in juveniles it is yellowish with green rays. Ligament is located posterior of beaks and is well developed. Nacre is rose or yellowish olive centrally and bluish white or blue at margins. Dentition is rudimentary. Typically there is one small, poorly developed pseudocardinal in each valve. Lateral teeth are vestigial if present at all. Foot and muscles tend to be orange colored. **DISTRIBUTION:** Formerly known from the North Fork of Shenandoah River, the Potomac River, Bull Run, and the Calfpasture River, this species has become extremely rare. It is known recently only from the upper North Fork of Shenandoah River and Broad Run, Prince William County. **HABITAT:** The brook floater typically occurs in smaller creeks associated with rocky or gravelly substrates, especially in and near riffles and rapids. It does not occur in impoundments. Life history: A long-term brooder, gravid females have been recorded from late August to May. Glochidia are released in the spring. Host fish unknown. **THREATS:** Poor water quality seems to be the main threat. This species does not seem to tolerate silt or nutrient pollution well. Impoundments also eliminate habitat. The brook floater is surviving notably better north of Virginia. **NOTES:** In Virginia, the species has been recorded only once outside of the Potomac River basin. This is unusual in that the brook floater is recorded widely north of Virginia and also is known from several river systems south of Virginia.

Prepared by Philip H. Stephenson, Richmond, VA, and Sue A. Bruenderman, Missouri Department of Conservation, May 1994.

Slippershell Mussel

Alasmidonta viridis

ENDANGERED

DESCRIPTION: One of the smallest freshwater mussels, the slippershell does not exceed 2 inches in length. The anterior end is rounded, the posterior squared or truncated with a high, rounded posterior ridge. Beaks are elevated with five to six heavy loops running parallel to the growth lines. Valves are moderately inflated and often eroded in adults; epidermis is somewhat glossy and yellowish brown or light yellow with wavy green rays visible. The nacre is shiny and white or bluish white with a bluish tinge posteriorly. Distributon: Presently the slippershell occurs in Virginia only in the North Fork of Holston River and Big Moccasin Creek, Washington and Smyth counties, and Copper Creek and Little River, tributaries of the Clinch River in Scott and Russell counties. HABITAT: The slippershell has been found in substrates ranging from silts to cobbles and is often found in or around water willows, *Justicia americana*. It has been found in slow to moderate currents at depths up to 3 feet. It is found most frequently in headwaters of springfed streams. LIFE HISTORY: This species is a long-term brooder, with gravid females observed from fall through spring. The banded sculpin (*Cottus carolinae*), the johnny darter (*Etheostoma nigrum*), and the mottled sculpin (*Cottus bairdi*) have been reported as fish hosts. THREATS: Declining water quality has resulted in the loss of mussel populations and the extirpation of species such as the slippershell from the Powell River. Siltation and erosion from poor land-use practices are significant in the Holston River and its tributaries and are likely to have contributed to the decline of the slippershell. Pollution from toxic spills, inadequate wastewater treatment, and pesticide use account for the loss of populations in the North Fork of Holston River. Detrimental agricultural practices are affecting the populations of Big Moccasin Creek and Copper Creek. Dredging and channelizing are also major threats to this species. The impending invasion of the zebra mussel is a serious threat. NOTES: The slippershell mussel is widespread throughout the upper Mississippi River and Great Lakes drainages. This headwater species has always been rare in Virginia and is easily overlooked because of its small size.

Abridged from "Slippershell Mussel" by Lisie Kitchel, pp 266–267 in *Virginia's Endangered Species*, coordinated by Karen Terwilliger, 1991 (Blacksburg, VA: The McDonald & Woodward Publishing Company).

Tennessee Heelsplitter

Lasmigona holstonia

DESCRIPTION: The Tennessee heelsplitter, a relatively small mussel with a
length of about 2½ inches, is characterized by a smooth, elongate, rhomboid
shell that is moderately inflated and unsculptured. The shell is slightly thicker anteriorly
and has a sharply rounded anterior margin. The ventral margin is broadly rounded but
flattened centrally. The posterior margin is rounded below and truncated above. Beak is
slightly inflated and projects slightly above hinge line. Growth rings are shallow concentric grooves in which the epidermis is darker than the rest of the valve. Color of the
periostracum is variable, ranging from yellow to yellowish brown to blackish brown. The
nacre is white anteriorly and bluish white or iridescent posteriorly; salmon or yellow areas occur near the beak cavity. The nacre is thin along the margin of the shell and the
epidermis shows through. DISTRIBUTION: This headwater species is now only very rarely
encountered in Virginia. It is known from the North Fork of Holston River in Bland
County, the Middle Fork of Holston River in Smyth County, and the Clinch River in
Tazewell County. It is always in low numbers. HABITAT: The Tennessee heelsplitter occurs only in isolated populations in the extreme headwaters of rivers. LIFE HISTORY:
This primitive mussel is a long-term brooder that is gravid from September through May.
Host fishes are not known. THREATS: Pesticide residues from poor agricultural practices
and excessive silt and erosion are perhaps the greatest threats in the rivers containing
Tennessee heelsplitters in Virginia, but the loss of the species throughout its range was
undoubtedly also affected by industrial and municipal pollution. The impending invasion
of the zebra mussel is a seri-
ous threat. NOTES: The Tennessee heelsplitter is very rare
throughout its range in the
Tennessee and Alabama rivers and extremely rare in Virginia. Its small size and rarity
make it difficult to find. The
creation of mussel sanctuaries
or conservation zones to limit
agricultural and development
activities along reaches of the
North and Middle forks of
Holston River and the Clinch
River would provide protection for this and other rare
headwater species.

Abridged from "Tennessee Heelsplitter" by Lisie Kitchel, pp. 267–269 in *Virginia's Endangered Species*, coordinated by Karen
Terwilliger, 1991 (Blacksburg, VA: The McDonald & Woodward Publishing Company).

Little-Wing Pearlymussel

Pegias fabula

ENDANGERED

DESCRIPTION: The little-wing pearly-mussel is a small freshwater mussel reaching a maximum length of about 1½ inches. Live specimens are usually eroded, which gives them a chalky or ashy white appearance. The epidermis, when present, is light green or dark yellow brown, with broad to narrow dark rays along the anterior portion of the shell. Valves are thickened anteriorly. The nacre is whitish on the anterior border and usually salmon or flesh-colored in the beak cavity. Valves are sexually dimorphic: females have an obvious posterior ridge and truncated posterior end. DISTRIBUTION: This species occurs in the Clinch and North Fork of Holston rivers. Recent specimens have come from the upper Clinch River at Cliffield, Tazewell County; the Little River, Russell County; and in North Fork of Holston River between Saltville and Nebo, Smyth County. HABITAT: The little-wing pearlymussel is a lotic, riffle-dwelling species that is usually found in the headwaters of high-gradient tributary streams. Individuals have been found in the transition zone between pools and riffles, buried under large flat rocks, and in gravel substrates adjacent to water willow, *Justicia americana*, beds. LIFE HISTORY: This is a winter or long-term brooder which retains glochidia from midsummer to spring of the following year. Fish hosts are unknown, but likely hosts include the banded sculpin (*Cottus carolinae*) and redline darter (*Etheostoma rufilineatum*). THREATS: Given the small size of the upper North Fork and its remoteness from urban development, potential threats appear limited to increased logging, oil and gas exploration, and overcollecting. The extreme rarity of this species in the headwaters of the upper Clinch River in Tazewell County suggests that only a remnant population exists there and it may be on the verge of extirpation. Inadequate treatment of sewage has been reported as responsible for the decline of the little-wing pearly-mussel in the Clinch River. The impending invasion of the zebra mussel is a serious threat. NOTES: The little-wing pearlymussel is very rare throughout its range in the upper Tennessee and Cumberland drainages and extremely rare in Virginia. The upper reaches of the North Fork of Holston and Clinch rivers must remain pristine if the little-wing pearlymussel is to survive in Virginia.

Abridged from "Little-Wing Pearlymussel" by Steven A. Ahlstedt, pp. 270–271 in *Virginia's Endangered Species*, coordinated by Karen Terwilliger, 1991 (Blacksburg, VA: The McDonald & Woodward Publishing Company).

Elephant-Ear
Elliptio crassidens

ENDANGERED

DESCRIPTION: Valves are heavy, slightly inflated, and oblong in shape, measuring up to 5 inches. The posterior ridge is prominent, ending as a blunt point at the posterior end. The epidermis is brown to black, and the surface is marked with prominent growth lines. This species is distinguished by its heavy, long, dark-colored and somewhat flat-sided shell; prominent posterior ridge; and pink to purple nacre. DISTRIBUTION: The elephant-ear is reported from only a few localities in two rivers: the Clinch River in Scott County near Dungannon and Clinchport, and the Powell River in Lee County at Fletcher Ford. HABITAT: This species is typically found in large rivers. It is usually associated with coarse sand and gravel substrates in rivers that are at least 6 feet deep and have strong currents. LIFE HISTORY: The elephant-ear is a short-term brooder. Gravid females have been recorded from April through July. The skipjack herring, *Alosa chrysochloris*, is apparently a fish host. THREATS: Degradation of water quality is the major threat to this species in the lower Clinch and Powell rivers. Cumulative effects of increased sedimentation in the Powell River from coal mining activities and other point and non-point sources of pollution have resulted in a reduction of the range of this species. NOTES: The elephant-ear is common locally throughout the Mississippi River drainage and several southern rivers, but it is extremely rare and of peripheral occurrence in Virginia. If the existence of the elephant-ear in Virginia is to continue, immediate measures must be taken to improve water quality in the Clinch and Powell rivers. Continued inventory and research should be conducted to determine the extent and viability of any remnant populations.

Abridged from "Elephant-Ear" by Michael L. Lipford, pp. 271–272 in *Virginia's Endangered Species*, coordinated by Karen Terwilliger, 1991 (Blacksburg, VA: The McDonald & Woodward Publishing Company).

Shiny Pigtoe

Fusconaia cor

ENDANGERED

DESCRIPTION: This is a medium-sized freshwater mussel reaching a maximum length of about 3 inches. It is distinguished by a smooth, shiny epidermis with prominent, wide dark green rays on a yellow to brown background. Valves are nearly triangular with surfaces marked by concentric growth rings and a median indentation. The beak cavity is deep. Nacre is white. Valves do not exhibit sexual dimorphism. DISTRIBUTION: This species is known in Virginia from the North Fork of Holston River above Saltville, Smyth County; the Clinch River from the Virginia-Tennessee border upstream to Nash Ford, Russell County; Copper Creek, Scott County; and Powell River from the Virginia-Tennessee border upstream to Hurricane Bridge, Lee County. HABITAT: The shiny pigtoe is a lotic species, occurring at fords, shoals, and other relatively shallow areas with moderate to fast currents. It is typically well-burrowed in stable substrates ranging from sand to cobbles. LIFE HISTORY: The shiny pigtoe is a short-term brooder that spawns in late May to early June and is gravid from mid-June to mid-July. The following fish species may be hosts in the upper North Fork of Holston River: whitetail shiner (*Notropis galacturus*), common shiner (*Notropis cornutus*), warpaint shiner (*Notropis coccogenis*) and telescope shiner (*Notropis telescopus*). Longevity of this species is about 24 years. THREATS: The present population in the Powell River, Lee County, may be threatened by oil and gas drilling and by the impact of coal mining on water quality and the river substrate. Coal mining wastes and run-off from Wise County appear to adversely affect downstream habitats and biota. Populations in the North Fork of Holston and Clinch rivers were reduced by toxic discharges and spills prior to 1972 and have not recovered. Environmental conditions in the North Fork of Holston and Clinch rivers appear satisfactory, but invasion of the Asian clam may prove detrimental. The impending invasion of the zebra mussel is a serious threat. NOTES: The shiny pigtoe is rare throughout its range in the upper Tennessee River drainage. Specimens from lower portions of rivers are somewhat heavier, more inflated, and usually dark in comparison with those from headwaters, which are more compressed and lighter in color.

Abridged from "Shiny Pigtoe" by Richard J. Neves, pp. 272–274 in *Virginia's Endangered Species*, coordinated by Karen Terwilliger, 1991 (Blacksburg, VA: The McDonald & Woodward Publishing Company).

Fine-Rayed Pigtoe

Fusconaia cuneolus

ENDANGERED

DESCRIPTION: This is a medium-sized freshwater mussel reaching a maximum length of about 3 inches. The species is distinguished by its satin-like epidermis with fine green rays on a yellow to brown background. Valves are nearly triangular in outline with surfaces marked by indistinct growth rings and a median, shallow indentation from the valve margin toward the beak. Anterior ends of the valves are rounded; ventral margins are almost straight. The beak cavity is moderately deep. Nacre is white. Valves exhibit no sexual dimorphism. The fine-rayed pigtoe and shiny pigtoe are similar in appearance; texture of the epidermis and ray patterns differentiate them. **DISTRIBUTION:** Known in Virginia in the Clinch River from the Virginia-Tennessee border to Cedar Bluff, Tazewell County; Copper Creek, Scott County; Little River, Russell County; and Powell River, Lee County. **HABITAT:** This is a lotic, riffle-dwelling species that usually inhabits ford and shoal areas of rivers with moderate gradient. It is typically well burrowed in stable substrates of mixed particle sizes ranging from sands to cobbles. **LIFE HISTORY:** Fine-rayed pigtoes are short-term brooders that spawn in May and are gravid until late July. The following fish species are known to be hosts: river chub (*Nocomis micropogon*), whitetail shiner (*Cyprinella galactura*), central stoneroller (*Campostoma anomalum*), telescope shiner (*Notropis telescopus*), Tennessee shiner (*Notropis leuciodus*), white shiner (*Luxilus albeolus*), mottled sculpin (*Cottus bairdi*), and fathead minnow (*Pimephales promelas*). Longevity of the fine-rayed pigtoe is up to 35 years. **THREATS:** The remnant population in the Powell River, Lee County, may be threatened by oil and gas drilling, adverse impacts of coal mining on water quality, and coal waste deposition on the river bottom. Fecal coliform levels commonly exceed EPA standards because of inadequate or inefficient sewage treatment plants, particularly in the North Fork of Powell River. Water quality in most of the Clinch River appears fair to good, although elevated levels of some metals occur in the sediments at some sites. Invasion of the Asian clam in recent years may prove detrimental. Potential threats to the small populations in Copper Creek and Little River include deforestation, siltation, and other poor land-use practices. The impending invasion of the zebra mussel is a serious threat. **NOTES:** The largest population of this species resides in the Clinch River, Virginia. Future outlook for the species depends, therefore, on the stability and health of this population.

Abridged from "Fine-Rayed Pigtoe" by Richard J. Neves, pp. 274–275 in *Virginia's Endangered Species*, coordinated by Karen Terwilliger, 1991 (Blacksburg, VA: The McDonald & Woodward Publishing Company).

Atlantic Pigtoe

Fusconaia masoni

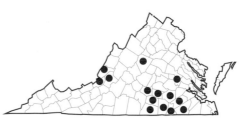

THREATENED

DESCRIPTION: Valves are medium-sized with lengths up to 2½ inches. Shape is rhomboidal, and the ventral margin is smoothly curved. The posterior ridge is broadly rounded and quite conspicuous. Valves are usually compressed, yet the beak still rises well above the dorsal margin. The beak is sculptured with three or four weak bars. The beak cavity is deep, the nacre usually white and somewhat shiny. Epidermis is often brownish yellow to greenish brown. **DISTRIBUTION:** The Atlantic pigtoe is known from the James, Roanoke, Meherrin, and Nottoway river systems. Recent records include specimens from tributaries of the James River between Albemarle and Henrico counties; the Craig Creek drainage in Craig, Allegheny, and Botetourt counties; the Appomattox River in Prince Edward County; the Nottoway River in Nottoway, Lunenburg, Dinwiddie, Brunswick, Greensville, Sussex and Southampton counties; and the Meherrin River in Mecklenburg County. **HABITAT:** This species seems to prefer clean, swift waters and is often found in gravel or sand and gravel substrates. **LIFE HISTORY:** The Atlantic pigtoe appears to be a short-term brooder. Eggs probably develop in May or June and are expelled before September. Host fishes are unknown. **THREATS:** Pollution, clearcutting, impoundments, and dredging are probably the major reasons for its reduced abundance and distribution in Virginia. The impending invasion of the zebra mussel is a serious threat.

NOTES: The species is uncommon throughout it range. The Nottoway River contains perhaps the largest population of the Atlantic pigtoe of any river and should be protected. The Atlantic pigtoe is often confused with the Virginia pigtoe, *Lexingtonia subplana*. These species have different beak sculptures, and the number and placement of their marsupial gills are quite different.

Abridged from "Atlantic Pigtoe" by Andy Gerberich, pp. 275–276 in *Virginia's Endangered Species*, coordinated by Karen Terwilliger, 1991 (Blacksburg, VA: The McDonald & Woodward Publishing Company).

Cracking Pearlymussel

Hemistena lata

ENDANGERED

DESCRIPTION: A medium-sized mussel reaching a maximum length of about 3½ inches, the cracking pearly-mussel is distinguished by thin, elongate, compressed valves; yellowish brown to greenish brown epidermis, often with dark green rays scattered over the surface; and a patch of purple in the beak cavity. Valves are rounded anteriorly and bluntly pointed posteriorly. Beak cavity is very shallow or absent, and most of the nacre is white to bluish iridescence. Valves do not meet symmetrically and often gape along the anterior end. DISTRIBUTION: In Virginia, this species occurs in the Powell River from the Virginia-Tennessee border upstream to Flanary Bridge, Lee County, and in the Clinch River from the Virginia-Tennessee border upstream to Fort Blackmore, Scott County. HABITAT: This is a lotic, riffle-dwelling species, occurring at fords and shoals with sand and gravel substrates and moderate current velocities. It can burrow deep into the river bottom because of an unusually long foot and is, therefore, difficult to collect. LIFE HISTORY: The cracking pearlymussel is apparently a short-term brooder. All other aspects of its life history are unknown. THREATS: The population in the Powell River may be threatened by oil and gas drilling, adverse impacts of coal mining on water quality, and coal waste deposition on the river bottom. An uncertain threat is the dispersal of the Asian clam into southwestern Virginia in the late 1970s; the impending invasion of the zebra mussel is considered a serious threat.

NOTES: This species also occurs in the Powell, Clinch, and Elk rivers in Tennessee and in the Green River in Kentucky. It has been extirpated from most of its former range; the largest remaining population occurs in the Clinch River about Norris Reservoir. The outlook for the species in Virginia is uncertain, but its survival in Virginia is dependent on the viability of the Clinch River population.

Abridged from "Cracking Pearlymussel" by Richard J. Neves, pp. 277–278 in *Virginia's Endangered Species*, coordinated by Karen Terwilliger, 1991 (Blacksburg, VA: The McDonald & Woodward Publishing Company).

Slabside Pearlymussel

Lexingtonia dolabelloides

THREATENED

DESCRIPTION: This is a medium-sized freshwater mussel reaching a maximum length of about 3½ inches. It is distinguished by a somewhat triangular outline, prominent beaks with shallow cavities located near the anterior end, solid valves with tawny to brown epidermis and broken green rays or blotches, and white to straw colored nacre. The ventral margin and dorsal slope of valves are curved. The epidermis usually contains prominent growth lines. If uneroded, the beak is sculpted with 6 to 8 crowded, wavy bars that are distinct anteriorly. Shell characters of the slabside are very similar to those of *Pleurobema oviforme*; consequently these species are best distinguished by characters of the soft parts. **DISTRIBUTION:** The slabside is known to occur in the Middle Fork of Holston River near Chilhowie, Smyth County; North Fork of Holston River above Saltville, Smyth County; Clinch River from the Virginia Tennessee border upstream to Fort Blackmore, Scott County; and Powell River below Jonesville, Lee County. **HABITAT:** The slabside pearly mussel resides in shoal and riffle habitats of intermediate sized streams characterized by moderate to fast-flowing water and a clean, heterogeneous substrate. **LIFE HISTORY:** The slabside is a short-term brooder with females gravid between mid-May and early August. Six species of minnows have been implicated as likely fish hosts: popeye shiner (*Notropis ariommus*), rosyface shiner (*Notropis rubellus*), saffron shiner (*Notropis rubricroceus*), silver shiner (*Notropis photogenis*), telescope shiner (*Notropis telescopus*), and Tennessee shiner (*Notropis leuciodus*). **THREATS:** The most secure population in Virginia occurs in the upper North Fork of Holston River, where subtle but long-term threats such as pollution and urbanization eventually may threaten it. The historic population in the river below Saltville was eliminated by mercury and chloride wastes from the now defunct alkali plant at Saltville. Remnant populations in the Clinch and Powell rivers may not be viable because of low numbers and isolated occurrence at only a few sites. Of unknown consequence is the impact of the recent invasion of the Asian clam. The impending invasion of the zebra mussel is a serious threat. **NOTES:** The slabside pearlymussel is rare throughout its range in the upper Tennessee River drainage. Impoundments, siltation, and water pollution likely have contributed to the extirpation of this species from the mainstem Tennessee River and have relegated extant populations to less disturbed river reaches.

Abridged from "Slabside Pearlymussel" by Richard J. Neves, pp. 278–279 in *Virginia's Endangered Species*, coordinated by Karen Terwilliger, 1991 (Blacksburg, VA: The McDonald & Woodward Publishing Company).

Sheepnose

Plethobasus cyphyus

DESCRIPTION: Valves are oval, somewhat elongated, inflated, and thick; the anterior end is rounded, the posterior end is long and bluntly pointed. Measuring up to 4¾ inches in length, valves are distinctly longer than high. Usually a row of large, broad, knob-like projections is evident on the center of each valve extending from the beak to the ventral margin. In older adults, the epidermis is yellowish to dark brown, whereas in younger specimens it is more yellow. Slightly iridescent posteriorly, the nacre is most often white; the color of the soft parts is distinctly orange. **DISTRIBUTION:** This species is of peripheral occurrence in Virginia, where it is known from two river reaches: the Clinch River from near Dungannon, Scott County, to the Virginia-Tennessee border, and the Powell River from Flanary Bridge, Lee County, to the Virginia-Tennessee line. **HABITAT:** This is a species found in large rivers and absent from small tributary streams. Although generally associated with sand and gravel substrates of riffles and shoals, it also occurs in mud and sand habitat in areas of deeper water (over 6 feet). **LIFE HISTORY:** Little is known of the life history of the sheepnose. It is a short-term brooder, with gravid females reported from May through July. The sauger, *Stizostedion canadense*, has been reported as a fish host. **THREATS:** Cumulative factors leading to deterioration of water quality have resulted in the decline of many sensitive and peripheral species in Virginia, including the sheepnose. The impending invasion of the zebra mussel is a serious threat. **NOTES:** The sheepnose is found locally in the Mississippi River drainage. Over much of its range it has experienced a decline and, in many areas, it has been extirpated. Older individuals with eroded valves and poorly developed tubercles may resemble *Fusconaia subrofunda*.

Abridged from "Sheepnose" by Michael L. Lipford, pp. 279–281 in *Virginia's Endangered Species*, coordinated by Karen Terwilliger, 1991 (Blacksburg, VA: The McDonald & Woodward Publishing Company).

James Spinymussel

Pleurobema collina

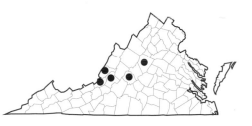

ENDANGERED

DESCRIPTION: The James spinymussel reaches a maximum length of a
little less than 3 inches. Adults are characterized by a dark brown epidermis
with prominent growth rings, a beak typically eroded and not elevated above the hinge
line, and the occasional presence of short bilateral spines on the dorsal surface of the
valves. Young specimens (less than 1½ inches) have a shiny yellowish epidermis with or
without one to three short spines. Valves are more or less rhomboidal to obvate in outline
with a broadly rounded anterior margin, straight to slightly concave ventral margin, and
straight to convex posterior end. Nacre color is white with occasional bluish suffusions.
The foot and mantle of the living specimen are conspicuously light orange in color. **DIS-
TRIBUTION:** This species is endemic to the James River drainage and is known from the
following streams: Potts Creek, Alleghany County; Craig, Johns, Dicks, Patterson, and
Catawba creeks in Craig and Botetourt counties; Meechums River, Moormans River,
and Rocky Run, Albemarle County; and Pedlar River, Amherst County. The prehistoric
range of the James spinymussel was the upper James River and its tributaries above the
Fall Line, but it is now restricted to small, headwater tributaries. **HABITAT:** This is a lotic
species that occurs in runs with moderate currents and sand, gravel, and cobble substrates.
Extirpated populations resided in sandy bottoms of larger streams with rather swift currents. Present populations occur in streams with water hardness values greater than 50
mg calcium carbonate per liter. **LIFE HISTORY:** The James spinymussel is a short-term
breeder, releasing glochidia in summer. The following fish hosts have been identified in
the Craig Creek drainage: rosyside dace (*Clinostomus funduloides*), bluehead chub (*Nocomis
leptocephalus*), mountain redbelly dace (*Phoxinus oreas*), blacknose dace (*Rhinichthys
atratulus*), central stoneroller (*Campostoma anomalum*), rosefin shiner (*Lythrurus ardens*),
satinfin shiner (*Cyprinella ostana*), and swallowtail shiner (*Notropis procne*). **THREATS:**
Potential threats to present populations include the upstream dispersal of the Asian clam,
poor logging or road con-
struction practices in the upper Craig Creek watershed in
Jefferson National Forest, and
sewage effluents from small
communities. The impending
invasion of the zebra mussel
is a serious threat. NOTES:
The James spinymussel is
confined to the upper James
River basin in Virginia and
West Virginia. It is extremely
rare throughout its range. The
decline of this species probably began with settlement
and industrialization of the
upper James watershed.

Abridged from "James Spinymussel" by Richard J. Neves, pp. 281–282 in *Virginia Endangered Species*, coordinated by Karen Terwilliger,
1991 (Blacksburg, VA: The McDonald & Woodward Publishing Company)

Ohio Pigtoe

Pleurobema cordatum

ENDANGERED

DESCRIPTION: The Ohio pigtoe is a medium-sized to large freshwater mussel with a maximum length of about 4½ inches. Valves are rather large and heavy with somewhat triangular outlines. Beaks are more or less elevated, swollen, and directed forward or incurved nearer to the anterior of the shell. Valves are swollen anteriorly, the diameter generally exceeding 50 percent of the length. Posteriorly, the shell is compressed with a broad more or less distinct radial furrow running from the beak toward the posterior margin and producing a shallow emargination. Surface of the shell is smooth. Epidermis may be light or dark brown, but generally is dark chestnut, becoming black on very old specimens. Growth lines are usually distinct and darker than the rest of the shell. Beak cavity can be deep but is often shallow. Nacre is white. Valves exhibit no sexual dimorphism. DISTRIBUTION: This species is known in Virginia only from the Clinch River at Pendleton Island, Scott County. HABITAT: The Ohio pigtoe is a lotic species residing in shoals of medium to large rivers with mud, sand, gravel, and cobble substrates. It has been collected in the Clinch River in the transition zone between pool and riffle. LIFE HISTORY: The Ohio pigtoe is a short-term brooder. Gravid females have been observed in April, May, June, and July. Reported fish hosts are the rosefin shiner, *Notropis ardens*, and bluegill, *Lepomis macrochirus*. THREATS: The remnant population in the Virginia portion of the Clinch River probably consists of only a few individuals. This population size is below the number required for successful reproduction and for this reason likely will be extirpated. The impending invasion of the zebra mussel is a serious threat. NOTES: The Ohio pigtoe is a common and commercially valuable species with a widespread distribution throughout the Mississippi, Ohio, Tennessee, and Cumberland rivers; but it is extremely rare in Virginia, where it exists on the margin of its range. A small reproducing population of the Ohio pigtoe occurs in the Tennessee portion of the Clinch River downstream from the Virginia-Tennessee border. Construction of Norris Dam and Reservoir divided the original population, which was relatively large historically in the Tennessee portion of the river.

Abridged from "Ohio Pigtoe" by Steven A. Ahlstedt, pp. 282–283 in *Virginia's Endangered Species,* coordinated by Karen Terwilliger, 1991 (Blacksburg, VA: The McDonald & Woodward Publishing Company).

Rough Pigtoe

Pleurobema plenum

ENDANGERED

DESCRIPTION: The rough pigtoe is a medium-sized freshwater mussel which reaches a maximum length of about 2½ inches. It is distinguished by a brown, satin-like epidermis; a beak above the hinge line turned anteriorly; and a radial depression in the central portion of the valves. Valves are solid and nearly triangular in outline. The ventral margin is nearly straight. The epidermis has irregular, concentric growth rings and usually lacks rays except on young specimens. Beak cavities are deep, and nacre color is white or pink. DISTRIBUTION: There are no recent records of this species within the state. Since there is a population nearby in Tennessee, it may occur in the lower Clinch River in Virginia, but it has not been documented. HABITAT: The rough pigtoe is a lotic species, residing in shoals of medium to large rivers with sand and gravel substrates. It has been collected in the Clinch River in the transition zone between pool and riffle and in sandy substrates. LIFE HISTORY: The rough pigtoe is apparently a short-term brooder. Fish hosts are unknown, as are other aspects of its life history. THREATS: A decline in water quality in the Clinch River basin could accompany the urban and rural development that appears likely to increase in that watershed. The impending invasion of the zebra mussel is a serious threat. NOTES: The rough pigtoe is typically rare throughout most of its range in the Tennessee and Cumberland rivers. It was listed as federally endangered in 1976. The remnant population in the Clinch River above Norris Reservoir, Tennessee, consists of few individuals and may soon be extirpated. The species probably is extirpated from Virginia.

Abridged from "Rough Pigtoe" by Richard J. Neves, pp. 284–285 in *Virginia's Endangered Species*, coordinated by Karen Terwilliger, 1991 (Blacksburg, VA: The McDonald & Woodward Publishing Company).

Pink Pigtoe

Pleurobema rubrum

DESCRIPTION: The pink pigtoe is a medium-sized to large freshwater mussel which reaches a maximum length of approximately 4½ inches. The shell is very oblique, having the outline of a right-angle triangle. Beaks are high and full, solid, inflated, and turned forward. The posterior ridge is low, rounded, and ends in a rounded point at the base of the shell. A median ridge is very high, rounded, curved, and usually separated from the posterior ridge by a radial, concave depression. The surface of the shell has irregular growth lines. Epidermis varies from brown to blackish to brownish green; rays are present on younger specimens. Beak cavities are deep and compressed; nacre is rose-colored or white, rarely yellowish or salmon. Sexual dimorphism is unknown. **DISTRIBUTION:** The species is known in Virginia only from the upper Clinch River at Pendleton Island, Scott County. **HABITAT:** The pink pigtoe is a lotic species, residing in shoals of medium to large rivers with mud, sand, gravel and cobble substrates. It has been collected in the Clinch River in gravel and cobble riffles and in the transition zone between pool and riffle. **LIFE HISTORY:** This species is a short-term brooder. Gravid females have been observed in May and July. Fish hosts are unknown. **THREATS:** The remnant population in the Virginia portion of the Clinch River probably consists of only a few individuals and likely will be extirpated because the population is too small to reproduce. The impending invasion of the zebra mussel is a serious threat. **NOTES:** The pink pigtoe is uncommon throughout its range in the middle Mississippi River drainage and extremely rare in Virginia. A small reproducing population occurs in the Tennessee portion of the Clinch River approximately 20 miles downstream from the Virginia-Tennessee border. Construction of Norris Dam and Reservoir probably divided the original population, which historically occurred throughout the Tennessee portion of the river.

Abridged from "Pink Pigtoe" by Steven A. Ahlstedt, pp. 285–286 in *Virginia's Endangered Species*, coordinated by Karen Terwilliger, 1991 (Blacksburg, VA: The McDonald & Woodward Publishing Company).

Rough Rabbitsfoot

Quadrula cylindrica strigillata

THREATENED

DESCRIPTION: The rough rabbitsfoot is one of the most unusual mussels found in Virginia. Valves are elongate and cylindrical in form, approximately three times longer than high. A row of knobs extends over an inflated posterior ridge with wrinkles on the dorsal ridge. The epidermis is usually marked by a pattern of dark green chevrons. Beaks are full and elevated, and beak sculpture consists of a series of coarse folds or wrinkles ending or continuing on the beak as small tubercles. The beak cavities are deep and compressed. Growth lines usually produce low ridges and grooves, darkly pigmented at major rest periods. The foot can be white but is often orange with a black striped background. Nacre is silvery white, iridescent, and thinner posteriorly, giving the valve an irregular thickness. **DISTRIBUTION:** Historically found in the Clinch, Powell, and Holston rivers, the rough rabbitsfoot still occurs in the Clinch River, Scott and Tazewell counties; Copper Creek, Scott County; Powell River, Lee County; and North Fork of Holston River, Washington County. It is usually rare in the Clinch and Powell rivers. **HABITAT:** The rough rabbitsfoot occurs only in the small headwater tributaries of the Tennessee River, often near the banks, in shoals with clean water and gravel bottoms or in riffles of shallow water. **LIFE HISTORY:** This subspecies is a summer brooder, with gravid females occurring in May through early July. Three species of fish have been identified as hosts: whitetail shiner (*Notropis galacturus*), spotfin shiner (*Notropis spilopterus*), and bigeye chub (*Hybopsis amblops*). **THREATS:** As with all freshwater mussels, the greatest threats to the rough rabbitsfoot are loss of river habitat and declining water quality.

Pollution from domestic sewage, small industrial plants, and mine waste run-off causes changes in water chemistry; these changes as well as toxic spills have accounted, in part, for the decline of the rough rabbitsfoot. The impending invasion of the zebra mussel is a serious threat. **NOTES:** The rough rabbitsfoot is widespread but uncommon in the Tennessee River system and localized in occurrence in Virginia, where populations are declining.

Abridged from "Rough Rabbitsfoot" by Lisie Kitchel, pp. 286–287 in *Virginia's Endangered Species*, coordinated by Karen Terwilliger, 1991 (Blacksburg, VA: The McDonald & Woodward Publishing Company).

Cumberland Monkeyface

Quadrula intermedia

ENDANGERED

DESCRIPTION: The Cumberland monkeyface is a medium-sized freshwater mussel reaching a maximum length of about 3¼ inches. It is distinguished by a deep radial depression along the posterior dorsal surface, many tubercles on the posterior two-thirds of the epidermis, and numerous green dots, chevrons, or zigzag lines on the yellow green epidermis. Valves are solid, relatively flat, and square to circular in shape with a notch on the posterior margin formed by the radial depression. The beak cavity is deep, and the nacre is generally white. The Cumberland and Appalachian monkeyfaces are very similar in appearance and easily confused. **DISTRIBUTION:** The Cumberland monkeyface is known in Virginia only in the Powell River from the Virginia-Tennessee border upstream to White Shoals, Lee County. **HABITAT:** This is a lotic, fast water species, usually occurring in riffles and runs of small to mid-sized rivers. It has never been found in small streams or impounded portions of rivers. This species is typically well burrowed in stable substrates and occupies the same macrohabitats as the other endangered mussel species in the Powell River, Lee County. **LIFE HISTORY:** This species is a short-term brooder, spawning in spring and becoming gravid in May and June. The streamline chub, *Erimystax dissimilis*, and the blotched chub, *Erimystax insignis*, have been identified as fish hosts. **THREATS:** The Powell River population may be threatened by wastes from oil and gas drilling, by adverse impacts of coal mining on water quality, and by coal waste deposition in the river bottom. High fecal coliform levels are of concern in this river, particularly near communities along the North Fork of Powell River. The invasion of the Asian clam may prove detrimental to the species, and the impending invasion of the zebra mussel is a serious threat. **NOTES:** The Cumberland monkeyface is extremely rare throughout its range in the upper Tennessee River drainage. Its numbers are apparently declining in Virginia. If current trends continue, its extirpation is likely. The Cumberland monkeyface was listed as federally endangered in 1976.

Abridged from "Cumberland Monkeyface" by Richard J. Neves, pp. 287–288 in *Virginia's Endangered Species*, coordinated by Karen Terwilliger, 1991 (Blacksburg, VA: The McDonald & Woodward Publishing Company).

Pimpleback

Quadrula pustulosa pustulosa

THREATENED

DESCRIPTION: The common name pimpleback adequately describes the external appearance of this mussel. However, the shape and number of pustules are highly variable in this subspecies, and some immature and older specimens may lack pustules. In many specimens a wide green ray extends from the beak. Young individuals are generally squarish or truncated; mature individuals vary from truncated to circular in shape. The dorsal margin of the shell is nearly straight and short, while the ventral margin is rounded. Valves are solid and moderately inflated, with the posterior half thinner than the anterior. Pustules are restricted to the anterior portion of the shell; the remainder of the shell is inflated and smooth. The epidermis is yellowish green, brown, or black in older individuals; younger specimens are brownish or yellowish green. The nacre is white, but brown blemishes appear frequently in older valves. Beaks are high, swollen, and turned forward; beak cavities are deep. The pimpleback is most commonly confused with the purple wartyback, *Cyclonaias tuberculata*. These species are easily distinguished by nacre color, which is white in the pimpleback and purple in the purple wartyback. DISTRIBUTION: Historically found in the Clinch, Powell, and Holston rivers, the pimpleback is now very rare and restricted to the Clinch and Powell rivers of Lee, Scott, and Russell counties. HABITAT: The pimpleback has been found in substrates ranging in size from silt to cobble size. It has been reported in all stream habitats except those with shifting sand. It is found in medium-sized and large streams and has been collected in standing to swift water deeper than 4 inches. LIFE HISTORY: The pimpleback is short-term summer brooder. Fish species reported as hosts include channel catfish (*Ictalurus punctatus*), shovelnose sturgeon (*Scaphirhynchus platyrhynchus*), black bullhead (*Ictalurus melas*), brown bullhead (*Ictalurus nebulosus*), fathead catfish (*Pylodictis olivaris*), and white crappie (*Pomoxis annularis*). THREATS: As with all freshwater mussels, the greatest threats to the pimpleback are loss of stream habitat and decline in water quality. Erosion and siltation as a result of poor land-use practices and the application of pesticides in agricul-

tural areas have contributed significantly to the decline in abundance of the pimpleback. The impending invasion of the zebra mussel is a serious threat. NOTES: The pimpleback is widespread throughout large rivers of the Mississippi drainage and is at the eastern edge of its range in Virginia. Because the species was abundant and its valves are uniformly thick, it was highly valued for making buttons.

Abridged from "Pimpleback" by Lisie Kitchel, pp. 288–290 in *Virginia's Endangered Species*, coordinated by Karen Terwilliger, 1991 (Blacksburg, VA: The McDonald & Woodward Publishing Company).

Appalachian Monkeyface

Quadrula sparsa

DESCRIPTION: The Appalachian monkeyface is a medium-sized freshwater mussel reaching a maximum length of about 3 inches. It is distinguished by a radial depression in front of and behind the posterior ridge, surface covered only sparsely by tubercles anteriorly, and yellow green epidermis with small greenish chevrons or triangles. Valves are somewhat inflated, rhomboidal in outline, and have a shallow notch on the posterior margin formed by the radial depression. The beak cavity is deep, and nacre is generally white. DISTRIBUTION: The Appalachian monkeyface is known in Virginia from the Powell River from the Virginia-Tennessee border upstream to Flanary Bridge, Lee County; and from the Clinch River between Pendleton Island and Dungannon, Scott County. HABITAT: This is a lotic, fast water species that occurs in shallow riffles and runs. It resides in stable, silt-free areas with substrates of mixed particle size ranging from sand to cobble. LIFE HISTORY: This species is presumably a short-term brooder, as are other monkeyfaces, breeding from May to July. No other information is available on its life history. THREATS: The population in the Powell River may be currently threatened by oil and gas drilling, adverse impacts of coal mining on water quality, and coal waste deposition on the river bottom. Also of concern are high coliform levels from inadequate sewage treatment, particularly in communities along the North Fork of Powell River. The impending invasion of the zebra mussel is a serious threat. NOTES: The Appalachian monkeyface is rare throughout its range in the upper Tennessee River system and is extremely rare in Virginia. The population in the Clinch River is of questionable viability because numbers are low and occurrences are isolated. Its extirpation is likely within the next decade. The Appalachian monkeyface was listed as federally endangered in 1976.

Abridged from "Appalachian Monkeyface" by Richard J. Neves, pp. 290–291 in *Virginia's Endangered Species*, coordinated by Karen Terwilliger, 1991 (Blacksburg, VA: The McDonald & Woodward Publishing Company).

Fanshell

Cyprogenia stegaria

ENDANGERED

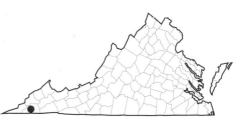

DESCRIPTION: Valves are solid, inflated, and somewhat rounded and reach a maximum length of about 2¼ inches. The posterior ridge is conspicuous and sharp dorsally, becoming less distinct and slightly double in older adults. The beak is curved upward and inward, only slightly elevated above the hinge line. The hinge ligament is black and conspicuous. Roughened with raised concentric growth lines, the epidermal surface exhibits one or two irregular lines of rounded knobs on the posterior two-thirds of the valve. The epidermis is greenish yellow to reddish brown, covered with small dark mottling that produces a ray pattern in most individuals. A white nacre is characteristic except posteriorly where it becomes iridescent. The rough epidermis with its strong concentric ridges and distinctive ray pattern distinguishes this species from others. **DISTRIBUTION:** The fanshell occurs in Virginia in the lower Clinch River in Scott County from Dungannon to the Virginia-Tennessee border. Its occurrence in Virginia is peripheral. **HABITAT:** The fanshell occurs in medium to large rivers and is associated with coarse sand and gravel substrates. It inhabits shoals and riffles of rivers with a strong current; and, in water 10 feet or deeper, it is associated with shoals and well-washed substrates. **LIFE HISTORY:** Little is known of the life history of this species. The species is probably a long-term brooder. Natural fish hosts are unknown. **THREATS:** The cumulative effects of increased sedimentation and water pollution have resulted in a reduction in population size. Degradation of water quality in the lower Clinch River poses the greatest threat to the continued existence of this species. The impending invasion of the zebra mussel is a serious threat. Notes. The fanshell is rare throughout its range in the Tennessee River drainage and Ohio River system. It is extremely rare in Virginia, where its population is declining. Unless immediate protection is provided, it may soon become extinct. The fanshell was listed as federally endangered in 1990.

Abridged from "Fanshell" by Michael L. Lipford, pp. 291–292 in *Virginia's Endangered Species*, coordinated by Karen Terwilliger, 1991 (Blacksburg, VA: The McDonald & Woodward Publishing Company).

Dromedary Pearlymussel

Dromus dromas

ENDANGERED

DESCRIPTION: This is a medium-sized freshwater mussel reaching a maximum length of about 3½ inches. The species is distinguished by a round to nearly elliptical shape, tawny or brownish epidermis with numerous narrow rays of dots or broken lines mixed with wider green rays or blotches extending to the valve margin, and whitish pink nacre. Valves are solid, often with a curved row of small bumps externally near the midline extending to the ventral margin. Some of the concentric growth rings are raised or recessed on the valves. Beak cavities are deep. Valves exhibit no known sexual dimorphism. **DISTRIBUTION:** The dromedary pearlymussel inhabits the Powell River from the Virginia-Tennessee border upstream to White Shoals, Lee County, and the Clinch River between Slant and Fort Blackmore, Scott County. **HABITAT:** The dromedary pearlymussel is a lotic, riffle-dwelling species that usually inhabits shoals and fords with moderate current velocities. However, this species does occur in deeper, slow-moving waters in Tennessee. It is typically well burrowed in silt-free, stable substrates of mixed particle sizes ranging from sand to cobble. **LIFE HISTORY:** The dromedary pearlymussel apparently is a long-term brooder because gravid females have been reported in April and October Fish hosts are unknown, although the gilt darter (*Percina evides*) is a possible host. No other life history information is available. **THREATS:** The Powell River population may be threatened by oil and gas drilling, by adverse impacts of coal mining on water quality, and by coal waste deposition on the river bottom. Also of concern are high coliform levels resulting from inadequate sewage treatment, particularly in communities along the North Fork of Powell River. Invasion of the Asian clam is problematic; the impending invasion of the zebra mussel is a serious threat. **NOTES:** The dromedary pearlymussel is rare throughout its range in the Tennessee and Cumberland rivers and extremely rare in Virginia. The rarity and isolation of this species in the Clinch River of Virginia will likely lead to extirpation within the next decade if water quality does not improve. The dromedary pearlymussel was listed as federally endangered in 1976.

Abridged from "Dromedary Pearlymussel" by Richard J. Neves, pp. 292–294 in *Virginia's Endangered Species*, coordinated by Karen Terwilliger, 1991 (Blacksburg, VA: The McDonald & Woodward Publishing Company).

Cumberlandian Combshell

Epioblasma brevidens

ENDANGERED

DESCRIPTION: This medium-sized mussel (adult length 2 to 3 inches) exhibits pronounced sexual dimorphism. Valves are distinctly colored, with a yellowish periostracum covered with fine green broken rays sometimes appearing as dots. The sexes are separated by the shape of the valves. Male valves are nearly triangular or rhomboid and somewhat flat-sided, without pustules or ridges. Female valves are squarish with a greatly inflated posterior end. The marsupial swelling in females becomes sharply elevated and decidedly separated from the rest of the shell by two distinct grooves. The posterior margin of the female valve is toothed. The Cumberlandian combshell is distinct and does not closely resemble any other species in its genus. The distinctive ray pattern, which appears as rows of green dots or dashes, and the characteristic swelling on female valves identify this species. DISTRIBUTION: The Cumberlandian combshell has been reported in Virginia from the Powell River upstream to Rose Hill, Lee County; in the Clinch River upstream as far as Clinchport, Scott County; and in the North Fork of Holston River at Hilton, Scott County. HABITAT: The combshell is found with other riverine species in clean gravel shoals and riffles of medium-sized streams. LIFE HISTORY: This species is a long-term brooder. Gravid females have been reported in May and June. Reported fish hosts include the greenside darter (*Etheostoma blennioides*), spotted darter (*Etheostoma maculatum*), redline darter (*Etheostoma rufilineatum*), Tennessee snubnose darter (*Etheostoma simoterum*), logperch (*Percina caprodes*), and banded sculpin (*Cottus carolinae*). THREATS: This species appears to be declining for reasons not well understood. The greatest threats are deterioration of water quality and habitat alteration. The medium-sized streams which support this species are being subjected to increased siltation and pollution. The impending invasion of the zebra mussel is a serious threat.

NOTES: Members of the genus *Epioblasma* seem to be more sensitive to environmental change than are other species and are often the first to disappear from a faunal assemblage when environmental degradation occurs. The Cumberlandian combshell is extremely rare throughout its range in the Tennessee and Cumberland river systems. It was listed as endangered in Virginia in 1987.

Abridged from "Cumberlandian Combshell" by Sally Dennis, pp. 294–295 in *Virginia's Endangered Species*, coordinated by Karen Terwilliger, 1991 (Blacksburg, VA: The McDonald & Woodward Publishing Company).

Oyster Mussel

Epioblasma capsaeformis

ENDANGERED

DESCRIPTION: This species is small (adult length 2 to 3½ inches) and exhibits pronounced sexual dimorphism. It has a yellowish green epidermis with rather faint green rays. The nacre is bluish white to cream in color. Male valves are almost evenly elliptical and many have a double posterior ridge. Female valves have an enormously developed, rounded, compressed marsupial swelling which is dark green, thin, and occasionally toothed. The small size and green-rayed periostracum separate this species from most other similar species. The female is distinctive with her dark green posterior swelling. DISTRIBUTION: The oyster mussel was once widespread and locally abundant in the upper Tennessee River system in Virginia, including the Clinch, Powell, and North Fork of Holston drainages. It was the dominant member of the Speers Ferry mussel assemblage in 1973–1975. Recent surveys of the Clinch and Powell confirm the rapid decline of this species in these rivers. It is presently known from only the following sites in the Clinch River: Pendleton Island and Slant, Scott County; near Cleveland, Russell County; and above Cedar Bluff, Tazewell County. HABITAT: The oyster mussel commonly inhabited riffle and shoal areas of small to medium-sized streams. It was found in fine to coarse gravel and in pockets of gravel between bedrock ledges in areas of swift current. The oyster mussel also inhabited quieter shoal areas where substrates consisted of gravels and some mud. LIFE HISTORY: This species is a long-term brooder. Gravid females have been found during early spring, late summer, and fall. Fish species identified as hosts include the spotted darter (*Etheostoma maculatum*), redline darter (*Etheostoma rufilineatum*), dusky darter (*Percina sciera*), and banded sculpin (*Cottus carolinae*). THREATS: The recent decline in numbers of this species is most likely the result of subtle changes in the water quality of the medium-sized streams which it inhabits. The genus as a whole seems more sensitive to changes in habitat and water quality than other genera of freshwater mussel. Not enough is known about the habitat requirements, physiology, and life history of this species to determine the exact cause of its decline. For this reason, it may be difficult to prevent its extirpation from Virginia. NOTES: The oyster mussel is extremely rare throughout its range in the upper Tennessee River drainage and rapidly declining in Virginia, where it is on the verge of extirpation. Considering the low numbers of individuals that remain, conservation efforts may be too late to save it. This species was listed as endangered in Virginia in 1987.

Abridged from "Oyster Mussel" by Sally Dennis, pp. 295–296 in *Virginia's Endangered Species,* coordinated by Karen Terwilliger, 1991 (Blacksburg, VA: The McDonald & Woodward Publishing Company).

Tan Riffleshell

Epioblasma florentina walkeri

ENDANGERED

DESCRIPTION: This is mussel reaches
a maximum length of about 2½ inches.
Valves exhibit strong sexual dimorphism. The valves are rather thin and
elliptical in outline, much inflated in females, but only moderately so in males. The epi
dermis is honey-yellow with numerous rays. The anterior valve margin is projecting and
evenly rounded. The ventral margin is strongly convex in larger males, much less so in
females and smaller males. In females, the marsupial swelling is very pronounced and is
usually limited anteriorly and posteriorly by a deep, narrow groove. The tan riffleshell can
be distinguished from the closely related oyster mussel by its somewhat larger size, yellowish color, and slightly more inflated beak. Female tan riffleshells do not show the dark
green color on the marsupial swelling which is characteristic of the oyster mussel and
have more teeth on the posterior shell margin. DISTRIBUTION: The tan riffleshell has
been reported from the Middle Fork of Holston River, Smyth County; the South Fork of
Holston River, Washington County; and the Clinch River at Cedar Bluff, Tazewell County.
Its entire range appears now to be limited to one reach in the Middle Fork of Holston
River near Chilhowie, Smyth County, and the one site in the Clinch River. HABITAT:
Riffleshells are characteristically found in lotic habitats where they inhabit clean gravel
substrates. The tan riffleshell has been found in riffle and shoal areas of small to medium-
sized streams. LIFE HISTORY: Little is known of the life history of this rare subspecies.
It is probably a long-term brooder, releasing glochidia in late spring and summer. Fish
hosts are unknown. THREATS: The greatest threats are from channelization of the Middle
Fork of Holston River and industrial development in the towns of Marion and Chilhowie.
The impending invasion of the zebra mussel is a serious threat. NOTES: The tan riffleshell
is extremely rare throughout
its range in the upper Tennessee River drainage. It appears
to have been extirpated from
all its range outside of Virginia. The status of the populations in the Clinch and
Middle Fork of Holston rivers is presently unknown. Because biological studies may
impact this subspecies adversely, the best chance for its
recovery is habitat protection.
The tan riffleshell was federally listed as endangered in
1977.

Abridged from "Tan Riffleshell" by Sally Dennis, pp. 296–298 in *Virginia's Endangered Species*, coordinated by Karen Terwilliger,
1991 (Blacksburg, VA: The McDonald & Woodward Publishing Company).

Green Blossom

Epioblasma torulosa gubernaculum

ENDANGERED

DESCRIPTION: This is a medium-sized mussel (maximum length 3½ inches) which exhibits pronounced sexual dimorphism. Valves are generally compressed and solid, and the hinge ligament is short. The umbos are inflated with feeble sculpturing. The epidermis is smooth, shiny, and yellowish green with green rays. Male valves are generally irregularly ovate with a wide radial furrow ending in a broad sinus. Female valves are generally ovate, larger than the male, and have a large flattened, rounded marsupial swelling extending from the middle of the base to near the upper part of the posterior end. The male shows some signs of nodules along the posterior slope. **DISTRIBUTION:** In Virginia, this subspecies is reported from the North Fork of Holston River as far upstream as Holston Bridge, Scott County, and from the Clinch River as far upstream as Dungannon, Scott County. **HABITAT:** The green blossom is a riverine species which seems to require a clean gravel substrate. It is found in swift flowing water with riffles and shoals. **LIFE HISTORY:** The green blossom is probably a long-term brooder.

Fish hosts have not been identified. **THREATS:** If the green blossom remains in Virginia, it may inhabit a reach of the Clinch River. If it does, degradation of water quality and habitat would pose the major threats. **NOTES:** Taxonomic synonyms for this subspecies include *Dysnomia torulosa gubernaculum* and *Truncilla torulosa gubernaculum*. It was listed as federally endangered in 1976. It is likely that the green blossom is already extinct in Virginia.

Abridged from "Green Blossom" by Sally Dennis, pp. 298–299 in *Virginia's Endangered Species,* coordinated by Karen Terwilliger, 1991 (Blacksburg, VA: The McDonald & Woodward Publishing Company).

Snuffbox

Epioblasma triquetra

ENDANGERED

DESCRIPTION: This small species usually does not exceed 3 inches in length. Valves are triangular and inflated. The anterior end is rounded, and the posterior end truncated. The posterior ridge is sharply defined and forms a flattened posterior region. The epidermis is yellowish with many dark green broken rays which appear as rectangular or chevron-shaped blotches. Beak sculpturing is weak. Sexual dimorphism in valve shape is evident: the female valve is smaller than that of the male and much more inflated; it is more strongly and sharply truncate posterior. Edges of the female valves at the marsupial region are sometimes toothed and gape slightly when closed. The triangular shape, deep inflation, and characteristic chevron-shaped markings on the epidermis separate this species from most others that occur with it. Males may have a superficial resemblance to the deertoe, but the latter generally have a browner epidermis and are more compressed. The snuffbox is also similar in shape and color to the elktoe, *Alasmidonta marginata;* but the latter is much thinner-shelled and usually greener in appearance. DISTRIBUTION: The snuffbox is reported from the North Fork of Holston River at Mendota, Washington County, and in the Clinch River upstream as far as Clinchport in Scott County. It is rare in the Clinch River and the lowermost reaches of the Powell River in Lee County. HABITAT: The snuffbox prefers swiftly flowing water and is found with other riverine species in clean gravel shoals and riffles of medium-sized to large streams. LIFE HISTORY: The snuffbox is a long-term brooder. Gravid females have been found during all months except July. Reported fish hosts are the logperch (*Percina caprodes*) and the banded sculpin (*Cottus carolinae*). THREATS: The greatest threats are deterioration of water quality and habitat alteration. The medium-sized streams which now support this species are being subjected to increasing levels of siltation and pollution. The impending invasion of the zebra mussel is a serious threat. NOTES: The snuffbox is rare throughout its range in the Mississippi Basin and very rare in Virginia. Impoundment of large rivers has eliminated much of the habitat for the snuffbox. The snuffbox was listed as endangered in Virginia in 1987.

Abridged from "Snuffbox" by Sally Dennis, pp. 299–300 in *Virginia's Endangered Species*, coordinated by Karen Terwilliger, 1991 (Blacksburg, VA: The McDonald & Woodward Publishing Company).

Birdwing Pearlymussel

Lemiox rimosus

ENDANGERED

DESCRIPTION: The birdwing pearlymussel is small, with a maximum length
of about 2 inches. It is distinguished by
a dark green epidermis and corrugated
ridges on the posterior half of the valves. Valves are solid and triangular to ovate in outline. Beak cavities are rather deep; nacre is white. The species is sexually dimorphic: male
valves possess a shallow depression on the posterior slope, whereas female valves are more
ovate with a weakly developed swelling along the posterior ventral margin. The birdwing
is distinctive and easily identified. DISTRIBUTION: The birdwing pearlymussel is known
to occur in the following river reaches: Powell River from the Virginia-Tennessee border
upstream to Snodgrass Ford, Lee County; Clinch River from the Virginia-Tennessee border
upstream to Blackford, Russell County; and lower Copper Creek, Scott County. HABI-
TAT: This is a lotic, riffle-dwelling species that usually occurs in moderate- to fast-flowing water of shallow to moderate (6 feet) depth. It resides in stable, silt-free substrates of
mixed particle sizes ranging from sand to cobble. LIFE HISTORY: The birdwing
pearlymussel is long-term brooder, becoming gravid in September and retaining glochidia
throughout the winter. The banded darter, *Etheostoma zonale*, and probably the greenside
darter, *Etheostoma blennioides*, are among its fish hosts. THREATS: Impoundments, siltation, and water pollution are likely factors contributing to the decline of this species.
The Powell River population may be threatened by oil and gas drilling, the adverse impacts of coal mining on water quality, and the deposition of coal wastes on the river bottom. Fecal coliform levels in the Powell River commonly exceed EPA standards because
sewage treatment is inadequate, especially in the North Fork. For similar reasons, water

quality in the upper Clinch
River is considered to be only
fair to good. The impending
invasion of the zebra mussel
is a serious threat. NOTES:
The birdwing pearlymussel is
rare throughout its range in
the upper Tennessee River
drainage and extremely rare in
Virginia. Population sizes
may have fallen below viable
levels in the Clinch and Powell drainages. The birdwing
pearlymussel was listed as
federally endangered in 1976.

Abridged from "Birdwing Pearlymussel" by Richard J. Neves, pp. 300–302 in *Virginia's Endangered Species*, coordinated by Karen
Terwilliger, 1991 (Blacksburg, VA: The McDonald & Woodward Publishing Company).

Fragile Papershell

Leptodea fragilis

ENDANGERED

DESCRIPTION: Shell is oblong, thin and compressed, and reaches a maximum length of 6 inches. The ventral margin is uniformly rounded. The anterior end is rounded; the posterior end is rounded in males and broadly expanded in females. Beak is flattened and only slightly elevated above hinge line. Beak sculpture consists of three or four faint double-looped bars. Epidermis is light yellow, often becoming a dirty yellow brown with age. Faint green rays may cover the shell, although some individuals are rayless. Beak cavity is shallow; nacre blue-white, pinkish dorsally, and highly iridescent throughout. DISTRIBUTION: The fragile papershell probably still occurs in the Clinch River. HABITAT: The fragile paper shell is found in diverse habitats including small streams as well as large rivers; in mud, and or gravel substrates, and in both clear and murky water. LIFE HISTORY: This species is a long-term brooder. Gravid females may be found from August through mid-July; glochidia are released in July and August. The freshwater drum, *Aplodinotus grunniens*, is the fish host. THREATS: As with all freshwater mussels, the greatest threat to the continued existence of the fragile papershell in Virginia is declining water quality and loss of stream habitat. The impending invasion of the zebra mussel is a serious threat. NOTES: Virginia is at the periphery of the fragile papershell's distribution area. Elsewhere in its range in the Mississippi River system it is one of the more common mussels found.

Based on information from "Fragile Paper Shell" by R. D. Oesch, pp. 172–175 in *Missouri Naiads, a guide to the mussels of Missouri* by R. D. Oesch, 1984 (Jefferson City: Missouri Department of Conservation); "Fragile papershell," pp. 120–121 in *Field Guide to Freshwater Mussels of the Midwest* by K. S. Cummings and C. A. Mayer, 1992 (Champaign: Illinois Natural History Survey); and R. J. Neves, pers. comm. 1994).

Black Sandshell

Ligumia recta

THREATENED

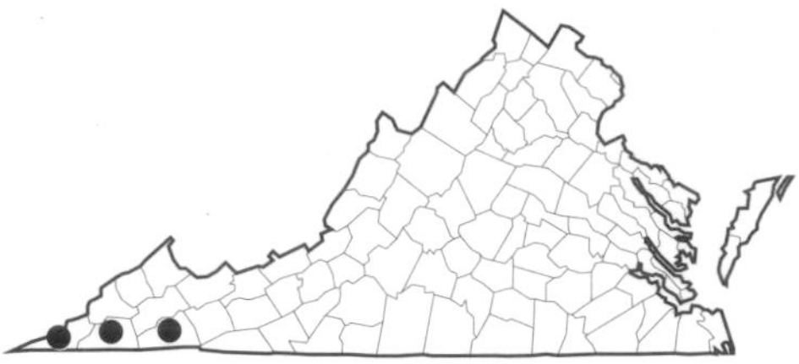

DESCRIPTION: Adult black sandshell commonly exceed 6 inches in length. Dorsal and ventral margins of the valves are almost parallel. Valves are more than twice as long as high and are thick and flattened. The beak is only slightly elevated above the hinge ligament. Sexual dimorphism is evident. The valves of males are pointed posteriorly; those of females are more evenly rounded and broader, particularly toward the posterior end. A black epidermis is typical in adults; younger specimens are greenish with pronounced rays. Except for growth rests, the epidermal surface is smooth. Nacre is typically white, but may be partially pink or purple, especially in the beak cavity and posteriorly. Hinge teeth are well developed and conspicuously elevated, one in the right valve and two in the left. The species is distinguished by its relatively large size, lanceolate shape, and black epidermis in adults. **DISTRIBUTION:** The occurrence of the black sandshell in Virginia is restricted to the lower reaches of four rivers: Middle Fork of Holston River in Washington County; North Fork of Holston River in Scott County near the Virginia-Tennessee border; Clinch River from Dungannon, Scott County, to the Virginia-Tennessee border; and Powell River in Lee County near the Virginia-Tennessee border. **HABITAT:** The black sandshell is usually found in medium to large rivers with a strong current. The species is closely associated with shoals of sand and gravel and only occasionally occurs in muddy substrates. It prefers water depths ranging from a few inches to 10 feet. **LIFE HISTORY:** The black sandshell is a long-term brooder. Gravid females have been reported during all months of the year. Fish hosts include bluegill (*Lepomis macrochirus*), white crappie (*Pomoxis anularis*), American eel (*Anguilla rostrata*), largemouth bass (*Micropterus salmoides*), and sauger (*Stizostedion canadense*). **THREATS:** Degradation of water quality poses the greatest threat to the continued existence of the black sandshell. The impending invasion of the zebra mussel is a serious threat. **NOTES:** The black sandshell is common in the Mississippi River drainage but very rare in Virginia. Historically, threats to the species included physical destruction of shoals and impoundments of rivers throughout its range.

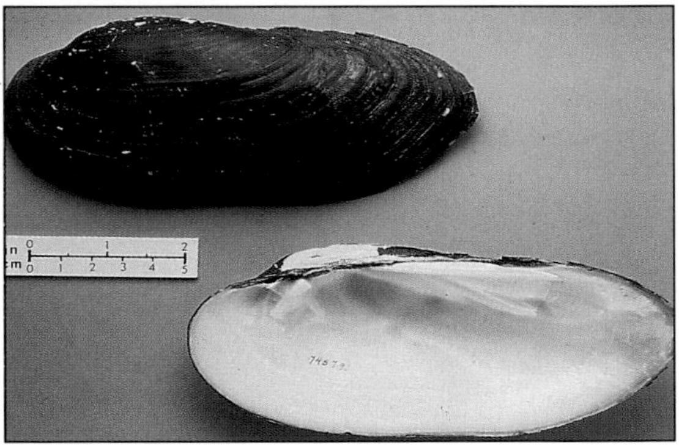

Abridged from "Black Sandshell" by Michael L. Lipford, pp. 302–303 in *Virginia's Endangered Species*, coordinated by Karen Terwilliger, 1991 (Blacksburg, VA: The McDonald & Woodward Publishing Company).

Purple Lilliput

Toxolasma lividus

ENDANGERED

DESCRIPTION: The purple lilliput is a small freshwater mussel with a maximum length of about 1¼ inches. It is distinguished by a black, blackish brown, or olive brown epidermis. The surface of the shell is rayless and coarse with small circular ridges. Valves are long and solid, elliptical or somewhat rhomboidal in shape. The posterior ridge is very faint, but the beak is relatively full and usually eroded. The beak cavity is shallow and consists of a few pits with muscle scars inside the shell. Nacre is generally a dark purple; there may be a white margin along the edge of the shell. Sexual dimorphism is exhibited: female valves are more rounded and truncate posteriorly than those of males. Gravid females have a fleshy white outgrowth on the mantle tissue. DISTRIBUTION: In Virginia, this species is known only from the Clinch River in Russell County and the North Fork of Holston River in Smyth County. HABITAT: The purple lilliput is typically a lotic species residing in headwaters of small to medium rivers. They have been collected in mud, sand, and gravel substrates. LIFE HISTORY: This species is a long-term brooder. Fish hosts include the green sunfish (*Lepomis cyanellus*) and the longear sunfish (*Lepomis megalotis*). THREATS: In the upper North Fork of Holston River, silt generated by agricultural land-use is probably the greatest threat to survival of mussel populations. The impending invasion of the zebra mussel is a serious threat. NOTES: Historically, the purple lilliput had a patchy distribution in Virginia. It is apparently extirpated from the Clinch River. The only known occurrence in Virginia is on one recent fresh-dead specimen from the North Fork of Holston River. The purple lilliput is relatively uncommon to rare throughout its range in the upper Tennessee and Cumberland rivers.

Abridged from "Purple Lilliput" by Steven A. Ahlstedt, pp. 303–304 in *Virginia's Endangered Species*, coordinated by Karen Terwilliger, 1991 (Blacksburg, VA: The McDonald & Woodward Publishing Company).

Deertoe

Truncilla truncata

ENDANGERED

DESCRIPTION: The deertoe is a small
to medium-sized mussel with a maximum length of about 2¾ inches. Valves
are triangular in shape, somewhat in-
flated, and generally exhibit a gape between the valves. The anterior end is rounded,
whereas the posterior end is bluntly pointed. The posterior slope is truncated and flattened, the reason for its specific name *truncata*. Generally dark green rays are joined to
form a striking zigzag pattern on a yellowish brown to green epidermal background. The
nacre is white, sometimes bluish and iridescent. Hinge teeth are well developed. The
triangular outline, pronounced posterior truncation, and unusual ray pattern are diagnostic for this species. DISTRIBUTION: Virginia is at the periphery of the deertoe's range.
The species occurs only in the Clinch River from Dungannon, Scott County, to the Virginia-Tennessee border and in the Powell River from Poteet Ford, Lee County, to the
Virginia-Tennessee line. HABITAT: The deertoe is characteristically found in medium to
large rivers and is associated with riffle and shoal areas in mud, sand, or gravel substrates.
LIFE HISTORY: The deertoe is long-term brooder. The freshwater drum, *Aplodinotus
grunniens,* and sauger, *Stizostedion canadense,* have been identified as fish hosts, but its
widespread occurrence in a variety of riverine habitats in the main parts of its range suggest that other species of fish also serve as suitable hosts. THREATS: Degradation of
water quality in the rivers supporting the deertoe poses the greatest threat to its survival.
The impending invasion of
the zebra mussel is a serious
threat. NOTES: The deertoe
is common in the Mississippi
River drainage but extremely
rare in Virginia, where its
numbers appear to be declining. It was one of seven species demonstrating a statis-
tically significant decline in
numbers from sites sampled
in 1979 and resampled in
1988. In the Clinch River, live
deertoes were found only at
Pendleton Island in 1988.

Abridged from "Deertoe" by Michael L. Lipford, pp. 304–305 in *Virginia's Endangered Species*, coordinated by Karen Terwilliger,
1991 (Blacksburg, VA: The McDonald & Woodward Publishing Comapny).

Purple Bean

Villosa perpurpurea

ENDANGERED

DESCRIPTION: The purple bean reaches a maximum length of about 2¼ inches. Valves are somewhat compressed and solid with beaks not very full or high. The surface of the valve has irregular growth lines, and color varies from brownish dark green to greenish back with narrow dark green, wavy rays dominant posteriorly. Beak cavities are shallow with two or three deep pits; muscle scars are noticeable inside the shell. Nacre ranges from deep purple to whitish purple, or purple with a blush of salmon. Valves exhibit sexual dimorphism. Male valves are straight or slightly concave on the ventral margin, whereas valves of females are more ovate and slightly emarginate below the faint marsupial swelling. The purple bean is often confused with the endangered Cumberland bean, *Villosa trabalis*. The two can be differentiated by the color of the nacre (purple versus white). DISTRIBUTION: The purple bean occurs in the Clinch River and Copper Creek, a tributary of the Clinch River. The largest of three populations occurs in the lower portion of Copper Creek in Scott County. HABITAT: The purple bean is a lotic, riffle-dwelling species that is restricted to headwater streams. It is found in moderate to fast-flowing water in clean-swept sand, gravel, and cobble substrates, and under large flat rocks. LIFE HISTORY: This species is a winter or long-term brooder. Fish hosts are unknown. THREATS: Given the small size of Copper Creek and its distance from urban development, identified threats are limited to silt from agricultural land-use and logging, oil and gas exploration, and the cutting of riparian vegetation along stream banks.

Copper Creek also has had elevated heavy metal concentrations. The impending invasion of the zebra mussel is a serious threat. NOTES: The purple bean is uncommon to rare throughout its range in the upper Tennessee River drainage. Once considered "not rare" in Virginia, only a small number of specimens have been reported in the past decade from the Clinch River between Cedar Bluff and Speers Ferry.

Abridged from "Purple Bean" by Steven A. Ahlstedt, pp. 305–306 in *Virginia's Endangered Species*, coordinated by Karen Terwilliger, 1991 (Blacksburg, VA: The McDonald & Woodward Publishing Company).

Unthanks Cave Snail

Holsingeria unthanksensis

ENDANGERED

DESCRIPTION: The unpigmented shells are about ¹/₁₆ to ³/₁₆ inch high and about ¹/₃₂ inch wide. There are 4½ to 5¼ whorls, with the body whorl 50 to 57 percent of the shell height. The aperture is nearly circular, expanded, and thickened all around. The operculum has an amber color and about four whorls. The dorsal surface is smooth, but the ventral surface has a horny, curving peg located at the end of the muscle scar attachment nearest the operculum nucleus. DISTRIBUTION: This snail has been reported only from Unthanks Cave in Lee County. HABITAT: The Unthanks Cave snail occurs on the undersides of small rocks in the stream of Unthanks Cave. LIFE HISTORY: No information is available on the life history of this species. THREATS: Any disruption of water flow or water quality in the underground stream of Unthanks Cave would be detrimental to the snail. NOTES: The Unthanks Cave snail is an extremely rare and isolated species known only from this one cave in Virginia. Studies of internal anatomy have revealed that it probably is most closely related to a similar unpigmented cave snail from southwest Texas. Seemingly similar cave snails exist in Europe, although they exhibit differences in characteristics of the operculum.

Abridged from "Unthanks Cave Snail" by Robert Batie, pp. 307–308 in *Virginia's Endangered Species*, coordinated by Karen Terwilliger, 1991 (Blacksburg, VA: The McDonald & Woodward Publishing Company).

Spiny Riversnail

Io fluvialis

THREATENED

DESCRIPTION: This large, gill-breathing freshwater snail reaches a maximum length of about 2 inches. Specimens are generally thick and solid and vary in color from tan to dark brown or olive green. Six or seven whorls are prominent on the exterior surface of the shell, and the epidermal surface is wrinkled. Shells exhibit no sexual dimorphism. The smooth form of this species exhibits folds or parallel ridges which are slightly nodulous but without pronounced spines. The spiny form has one to twelve pronounced, well-armored spines. **DISTRIBUTION:** The spiny riversnail is known to occur in the Clinch River from the Virginia-Tennessee border upstream to the mouth of the Little River, Russell County; and North Fork of Holston River along the Virginia-Tennessee border (Lee County) and at one site upstream of Saltville, Smyth County. **HABITAT:** The spiny riversnail is a lotic species, usually found on limestone bedrock ledges or rocky shoals in shallow, moderate to fast currents. Specimens are typically found attached to cobbles and boulders in the well-oxygenated water of shoals and rapids. They have been observed at depths up to 6 feet; however, most occur at depths of about 18 inches. **LIFE HISTORY:** Female spiny riversnails lay their eggs from late May to early July on surfaces of smooth rocks, shells, or coal chunks and, in many cases, on the insides of shells of freshly dead mussels. Females typically lay their eggs in an uneven line resembling a ribbon or in a spiral shape. The spiny riversnail is long-lived. Its food consists of the slimy algal coating and entangled organic debris on rocks. **THREATS:** The present population in the Powell River, Lee County, may be threatened by oil and gas exploration and the adverse impacts of coal mining on water quality and river substrates. High fecal coliform levels from inadequate sewage treatment along the North Fork of Powell River reduce water quality. **NOTES:** The spiny riversnail occurred in all major tributaries of the upper Tennessee River system. Impoundment has been the major factor contributing to the decline of this species. The spiny riversnail has been successfully reintroduced to the North Fork of Holston River. Other possible transplant sites within the historic range of the species should be sought.

Abridged from "Spiny Riversnail" by Steven A. Ahlstedt, pp. 308–309 in *Virginia's Endangered Species*, coordinated by Karen Terwilliger, 1991 (Blacksburg, VA: The McDonald & Woodward Publishing Company).

Virginia Fringed Mountain Snail

Polygriscus virginianus

ENDANGERED

DESCRIPTION: The shell of the Virginia fringed mountain snail is a disc
with a shallow umbilicus. The spire is
not prominent, and the dorsal and ven-
tral surfaces are nearly parallel. Roughly four whorls, regularly increasing, are evident on
the shell. A deep groove is present on the dorsal surface of the last third of the body
whorl. The shell has four prominent raised spiral lines (fringes) with less prominent spiral lines between them. The aperture is heart-shaped with a slightly reflected lip. It is the
only snail in Virginia to have the aperture lip detached fro the body whorl and deflected
toward the umbilicus side. Shell measures about $^3/_{16}$ inch in diameter and $^1/_{16}$ inch in
height. **DISTRIBUTION:** This snail is endemic to Virginia. Shells have been found only
along a six mile section of bluffs bordering the New River in Pulaski County, but living
snails have a documented range of only 250 feet along that river. They have been observed in the field in only two locations. Four surveys in the last 50 years have documented only 27 living snails. **HABITAT:** The Virginia fringed mountain snail is found in
association with fragmented Elbrook dolomitic limestone. Areas where snails have been
found are usually heavily shaded and may be overgrown with honeysuckle; they are free
from rich humus and are usually areas where moist limestone fragments are mixed with a
clay soil. This snail is a calciphile, and living individuals apparently occur in soil at depths
from 24 inches to as shallow as 4 inches. Live snails have never been observed or collected from the surface. **LIFE HISTORY:** No information is available on the life history of
this species. **THREATS:** Al-
most all of the living snails
observed in their natural
habitat were found in a single
rock pile beside a country
road. Any disturbance to this
rock pile, widening of the
roadway, or disturbance of the
road shoulder by spraying,
fire, or grading could cause
the extinction of this species.
NOTES: The Virginia fringed
mountain snail was listed as
federally endangered in 1978.
The common name "Virginia
coil" has been misapplied to
this species.

Abridged from "Virginia Fringed Mountain Snail" by Robert Batie, pp. 309–310 in *Virginia's Endangered Species*, coordinated by
Karen Terwilliger, 1991 (Blacksburg, VA: The McDonald & Woodward Publishing Company).

Rubble Coil

Helicodiscus lirellus

ENDANGERED

DESCRIPTION: The shell is disc-shaped with four and one half to five whorls. It is pale greenish yellow and dull, not glossy. The umbilicus is shallow but occupies about 45 percent of the shell width. The aperture is lunate, slightly reflected, and has two pairs of teeth on the outer and basal walls and one pair of teeth on the parietal wall. Shells are about ³/₁₆ inch wide by ¹/₁₆ inch high. DISTRIBUTION. The rubble coil is endemic to Virginia. It has been found only in Rockbridge County burrowed in shale rubble at the base of a hill. HABITAT: This species is a calciphile found in limestone-rich rubble at the base of a hill. LIFE HISTORY: No information is available on life history. THREATS: Any disturbance of the limestone talus where the snail occurs could cause the extinction of this species. NOTES: The rubble snail closely resembles the shaggy coil, which has hairs on its small ridges and larger apertural teeth.

Abridged from "Rubble Coil" by Robert Batie, pp. 310–311 in *Virginia's Endangered Species*, coordinated by Karen Terwilliger, 1991 (Blacksburg, VA: The McDonald & Woodward Publishing Company).

Shaggy Coil

Helicodiscus diadema

ENDANGERED

DESCRIPTION: The shell is a disc with a flattened or slightly depressed spire. It is dull greenish brown and opaque; shells have 4¼ to 5 whorls. The umbilicus is wide and shallow, showing all of the coils and occupying from 40 to 47 percent of the shell diameter. Whorls are rounded and slowly increasing, with the last whorl descending slightly. The shell is sculptured with coarse growth wrinkles and 11 to 13 pinched spiral threads which bear prominent curved hairs. The sutures are deeply impressed. Shell height is slightly less than ¹/₁₆ inch; diameter slightly less than ³/₁₆ inch. **DISTRIBUTION:** The shaggy coil is endemic to Virginia, found only in Alleghany and Rockbridge counties. In Alleghany County it is known from three locations, in Rockbridge County from only one. **HABITAT:** The shaggy coil has been collected in leaf litter at the base of limy shale outcrops and on thinly wooded limestone hills. It is abundant in the top layer of damp leaf litter on exposed limestone hillsides clad with locust scrub. Living specimens are quite rare in the deeper layers of leaves and soil where other species occur. **LIFE HISTORY:** No information is available on life history. **THREATS:** This species is restricted to leaf litter at lime-rich sites.

Removal of the woodlands, disturbance of the limestone hillsides, or disruptions of habitat that result in decreased leaf litter could cause the extinction of this species. **NOTES:** The apertural dentition is almost identical to that of *Helicodiscus multidens,* which occurs in nearby counties. The shaggy coil has fewer and coarser fringes on the body whorl and has large curved hairs on the lirae that are obvious to the naked eye.

Abridged from "Shaggy Coil" by Robert Batie, pp. 310–312 in *Virginia's Endangered Species,* coordinated by Karen Terwilliger, 1991 (Blacksburg, VA: The McDonald & Woodward Publishing Company).

Spirit Supercoil

Paravitrea hera

ENDANGERED

DESCRIPTION: Shells of the spirit supercoil are about ⅛ to ¼ inch high and ⁵/₁₆ inch wide, pale brown to white, and nearly glossy. Spire is low to conical with up to $7\frac{7}{10}$ whorls which gradually increase in size. The periphery of the body whorl is rounded; the nuclear whorl is smooth, but later whorls have regularly impressed growth lines that disappear underneath the rounded body whorl. The lip of the aperture is not reflexed. Aperture base is rounded and the opening is slightly wider than high, giving it a lunate shape. Shells are deeply umbilicate; the umbilicus being about one-seventh the diameter of the shell. Teeth may or may not be present in immature shells. DISTRIBUTION: The spirit supercoil is endemic to Virginia and has been reported from only one location in Pittsylvania County. HABITAT: The spirit supercoil has been found in leaf litter on stream banks. LIFE HISTORY: No information is available on life history. THREATS: Disturbance of wooded areas on river bluffs would reduce leaf litter and disrupt the habitat of this snail. Clearcutting, selective logging, or removal of the top layers of soil should be avoided. NOTES: Only shells of the spirit supercoil have been collected. Thorough surveys in Virginia and North Carolina are needed to determine the full range of this species.

Abridged from "Spirit Supercoil" by Robert Batie, pp. 312–313 in *Virginia's Endangered Species*, coordinated by Karen Terwilliger, 1991 (Blacksburg, VA: The McDonald & Woodward Publishing Company).

Brown Supercoil

Paravitrea septadens

DESCRIPTION: The shell is small, pale brown, and has a very low, nearly disc-shaped spire with shallow sutures between the whorls. There are 6 whorls which show gradual increase in size, with the last quarter of the body whorl being slightly expanded in mature shells. The periphery and the base of the shell are somewhat flattened. The umbilicus is deep, well-like, and about one-seventh of the shell diameter. Shell height is about 1/16 inch, diameter about 1/8 inch. DISTRIBUTION: The brown supercoil is endemic to Virginia and found only in Dickenson (five locations) and Buchanan (two locations) counties. HABITAT: The brown supercoil is usually found in pockets of deep, moist leaf litter on wooded hillsides at the base of hills and in ravines. LIFE HIS-TORY: No information is available on life history. THREATS: Any type of disruption of the wooded hillsides which would reduce leaf litter on the hills or in ravines would be detrimental to this species. NOTES: The brown supercoil is found in the same habitat as *Paravitrea multidentata* but is distinguished from it by a more rounded periphery, smaller umbilicus, and fewer and weaker growth lines. The lamellae are radial rather than oblique.

Abridged from "Brown Supercoil" by Robert Batie, p. 313 in *Virginia's Endangered Species,* coordinated by Karen Terwilliger, 1991 (Blacksburg, VA: The McDonald & Woodward Publishing Company).

Fishes

A rich and diverse freshwater fish fauna occurs within Virginia. Presently 210 species of freshwater fish are known from the state; when the fish are considered at the subspecific level, there are 230 taxa. Of the species, 192 are native, five others are probably native, and 13 clearly have been introduced. Eleven species are euryhaline (that is, they have a wide tolerance of salinity) and most normally range from freshwater to the ocean. Ten of the 11 euryhaline species are diadromous; they regularly migrate between freshwater and marine environments. Nine of the diadramous species are anadromous (they migrate from marine waters to spawn in freshwater), whereas the American eel is catadromous (it migrates from freshwater to breed in the ocean). The banded killifish, *Fundulus diaphanus*, is the only nonmigratory euryhaline species — it lives in both freshwater and brackish water. In this chapter freshwater and euryhaline species are considered to be freshwater fishes, and the aggregate is called the Virginia freshwater fish fauna or ichthyofauna.

Compared to the taxonomic diversity of freshwater ichthyofaunas of adjoining states, that of Virginia is moderately rich. It greatly exceeds the 107 species known from Maryland, but is much less than that of Tennessee, which with some 307 species has the richest North American fish fauna.

The freshwater fish fauna of Virginia has a complex evolutionary and dispersal history and is supported by highly varied geological and ecological conditions. Eight of the 10 major drainages in Virginia cross two to four of the provinces and have a great variety of aquatic habitats and species of fishes. The physical separation of drainages by physiographic divides and bodies of saline water is manifested by biological differences. The composition of the fish fauna of each drainage in Virginia is distinctive. Furthermore, the evolution of endemics, here defined as indigenous to a single drainage, is due to the isolation of drainages. Also contributing to the distinctiveness of the fish fauna of some drainages is the fact that the ranges of many freshwater fishes end in Virginia.

Adjacent drainages generally share more species of fish than do geographically distant drainages. Many similarities among fish faunas are attributed to biotic exchange following stream piracy in the past when a tributary of one drainage eroded headward and undercut one or more tributaries of another. Interdrainage dispersal of freshwater fishes also occurred when sea level dropped during glaciations and the lower ends of drainages now entering Chesapeake Bay formed a single freshwater drainage, essentially an enlarged Susquehanna River system. When the glaciers melted, sea level rose and these low-elevation rivers were "drowned" by the influx of brackish water or saltwater.

Fluviatile species, those that are adapted to one or more of the diverse habitats that occur in flowing waters, characterize the inland fish fauna of Virginia. Only two natural lakes exist in the state: Mountain Lake at 3,500 feet above sea level in the New River drainage and Lake Drummond near sea level in Dismal Swamp in extreme southeastern Virginia. These lakes differ tremendously, and each is depauperate in fish species.

The diversity of native Virginia fishes has declined by seven species since the "Master Naturalist" Edward Drinker Cope seined here in 1867. Abundance of populations and individuals has also declined as evidenced by the numbers of imperiled species. In a recent national list of fishes Virginia (tied with North Carolina) was ranked eighth among all states harboring the largest numbers of freshwater fishes imperiled on the state or national level.

The vast majority of vulnerable species are composed of very small individuals, but they are (or were) nonetheless important links in Virginia's natural ecological webs. The decline and disappearance of taxa indicates that modern aquatic systems in Virginia have reached levels of stress that are marginal or intolerable to many species. The rich inland fish fauna in Virginia is in trouble because the perturbations of ecosystems are chronic, cumulative, and catastrophic. Indeed, anthropogenic impacts on Virginia's aquatic systems are so extensive that probably *not one truly pristine stream remains in the state*. Although some species of fish are inherently more susceptible to disturbances than others, chronic stress on ecosystems often is mirrored by the decline of entire communities.

Factors contributing to the decline and demise of Virginia's freshwater fishes are essentially the same as those that are affecting adversely the ichthyofaunas of adjacent states. Probable causes include environmental degradation such as turbidity and siltation from sheet flow across cultivated land, woodland clearings, road and bridge construction, denuded banks, and riparian developments; channelization; impoundment; thermal alteration; water withdrawal; catastrophic chemical spills; and other forms of pollution from industrial, agricultural, and domestic sources.

Siltation, perhaps the most pervasive factor limiting freshwater fish populations of eastern North America, is responsible, in large part, for the decline of nearly every species treated in this chapter. Silt reduces or destroys habitat heterogeneity and primary productivity, increases egg and larval mortality, abrades organisms, and alters, degrades, and entombs macrobenthic communities. Influx of silt into a stream increases the quantity of fine particles suspended in the water and causes turbidity. Turbidity itself is a limiting factor, particularly to the many species of sight-feeding fishes. It also lowers the primary productivity of plants by limiting the penetration of sunlight into the water. Present measures to reduce soil erosion and the consequent movement of sediment into streams are inadequate. Silt is a silent, insidious, and pervasive agent of degradation to stream ecosystems. Silt does not cause catastrophic kills or produce foul odors, and often it is not overly obvious to the untrained eye, but nonetheless it degrades ecosystems.

The principal limiting factors causing reduction of euryhaline fish populations are pollution (which can be chronic and cumulative by the time river waters reach the estuary); siltation; blockage of access to spawning grounds by dams; and, for larger species, overfishing.

Most of the fishes now in jeopardy in Virginia, as well as those that are extinct or extirpated, tend or tended to inhabit medium-sized to large waterways. Fishes of large creeks[1] and rivers have been disproportionately more adversely affected by environmental degradation than have fishes of small creeks because larger waterways tend to carry more and varied chemical pollutants, some with synergistic effects, than do creeks. The moderate to low gradients of larger streams guarantee that they have lower rates of silt removal than most small creeks. Large streams are major sources of water; thus human populations are higher and industrialization is more extensive on their floodplains, and the channels are impounded more frequently.

Within a drainage, any given site on a middle- or higher-order stream tends to harbor more species of fish than occur in entire lower-order streams. On the whole, fishes that inhabit freely flowing waters have been reduced much more extensively than those that inhabit lakes and reservoirs. Of the listed species, only the blackbanded sunfish typically inhabits standing water in Virginia. Thus we view Virginia's rivers as critical "threads of life."

The fish fauna of the Roanoke River drainage has the highest number of endemic species and the most species in total among the Atlantic slope drainages considered herein. Eight species with restricted distributions, including four endemics, are vulnerable in the Roanoke River drainage.

Only one species, the endemic candy darter, *Etheostoma osburni*, is in trouble in the New River drainage. It is suspected to have disappeared recently from some major tributaries. Three other endemic species occur in the New River drainage, but these are not in danger.

The majority of species in Virginia that are in jeopardy or are no longer present occupy the Tennessee River drainage in the far southwestern part of the state. This circumstance reflects both the facts that the fish fauna of the Tennessee River drainage is naturally rich and that it has been subjected to many abuses. The North Fork of Holston River has a long history of chronic chemical stresses emanating from Saltville which resulted in major kills of fish; but judging from fish surveys, this waterway appears to have been recovering since the mid-1970s. The Clinch and Powell river systems host the most species-rich fish fauna in Virginia. Owing partly to its larger size, the Clinch River has yielded more species than the Powell River. Several species have recovered significantly or entirely from the vast alkaline waste spill of 1967 that killed essentially all fishes in the 75 miles of the Clinch River between Carbo and the Virginia-Tennessee boundary. The richest fish faunas in tributaries of the Clinch River are in the Little River and, particularly, Copper Creek.

The Big Sandy River system drains part of the Cumberland Plateau in southwestern Virginia. The endangered variegate darter and the extirpated blackside darter are or were known from the Big Sandy River drainage. Of all the drainages in Virginia, that of the Big Sandy has been the most extensively devastated by human activities (such as washing coal, which results in siltation of coal fines in

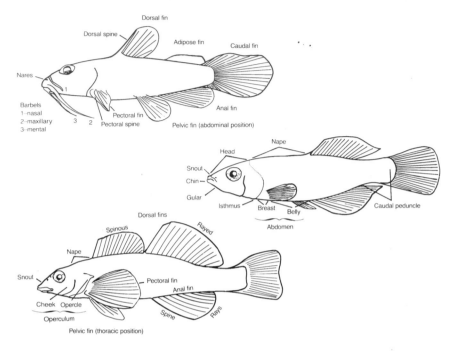

Figure 5. Generalized fishes showing selected morphological characters.

streams). The entire fauna of the Big Sandy drainage has been generally depleted in abundance and diversity by these activities.

The third decade of documenting intensively the condition of the Virginia fish fauna has begun. Yet, there is no evidence of substantial improvement in the circumstances of species of fishes in Virginia during the past 20 years other than that stemming from abatement of chronic chemical pollution in the North Fork of Holston and the Clinch river systems and the hiatus in massive kills of fish. Philosophically and practically, the general lack of recovery of Virginia's fish faunas during the last two decades reflects the scant emphasis on rehabilitating ecosystems compared to the more common practice of protecting species directly. The single most important requirement for rehabilitating aquatic ecosystems across Virginia is the long-term abatement of siltation.

Abridged from "Fishes" by Noel M. Burkhead and Robert E. Jenkins, pp. 321–337 in *Virginia's Endangered Species*, coordinated by Karen Terwilliger, 1991 (Blacksburg, VA: The McDonald & Woodward Publishing Company).

---

[1] The use of the terms "creek" and "river" can be vague and misleading. There is no fixed point, biologically speaking, at which a creek becomes a river. "Creek" is used generally for that subset of streams which range from tiny brooks to large streams with an average width of 65 feet. "River" is used for streams ranging in width from about 50 feet to more than a mile. The word "stream" is used in this chapter to refer to flowing waterways in general and includes both creeks and rivers.

Shortnose Sturgeon

Acipenser brevirostrum

ENDANGERED

DESCRIPTION: The shortnose sturgeon is small, with relative short snout,
wide mouth, posterior lower side lacking bony plates along and immediately
above anal fin base. Adults usually 18 to 36 inches long. The body is somewhat pentago-
nal in cross section and armored with five rows of bony plates (scutes). The length and
shape of the snout are variable and cannot always be used to identify individuals of the
species. DISTRIBUTION: Curiously only one certain record exists for the entire Chesapeake Bay basin; it is from the Potomac River, Washington, DC, in 1876. It is likely,
however, that the shortnose sturgeon was a resident of the Bay and occupied all four
major riverine estuaries of Virginia. HABITAT: Part of the population moves to spawning
grounds in large freshwater rivers in spring or fall. Spawning sites are swift water with
gravel and rubble substrates. Sturgeons winter in lower rivers and bays. LIFE HISTORY:
Shortnose sturgeon are benthic foragers. Juveniles feed on small crustaceans and insect
larvae. Adults in freshwater feed mostly on crustaceans, insect larvae, and mollusks; in
estuaries they mainly eat polychaete worms, crustaceans, and mollusks. This species is
anadromous. Individuals do not breed each year; the period between spawnings may be
ten years. THREATS: Pollution and siltation of riverine estuaries and the Chesapeake
Bay are the primary threats to the species. If it does reappear in the Chesapeake Bay
basin, harvesting by humans will be a threat also. Dams have probably blocked access to
spawning areas at or above the
Fall Line. NOTES: The shortnose sturgeon was among the
first animals given endangered status federally; this
occurred in 1967. The species
is considered extirpated in
Virginia. However, the
shortnose sturgeon has recently reappeared in the lower
Susquehanna drainage of the
upper Chesapeake basin, possibly entering via the Chesapeake and Delaware Canal, so
the potential for recolonization of Virginia's waters exists.

Abridged from "Shortnose Sturgeon" by Noel M. Burkhead and Robert E. Jenkins, pp. 342–343 in *Virginia's Endangered Species*,
coordinated by Karen Terwilliger, 1991 (Blacksburg, VA: The McDonald & Woodward Publishing Company).

Paddlefish

Polyodon spathula

THREATENED

DESCRIPTION: The paddlefish is large, often stout, with a paddle-shaped extension of the snout. Total length of adults ranges from 44 to 64 inches. The skin is smooth and nearly scaleless; head large, eyes small. There are two tiny barbels on the underside of the snout anterior to the mouth. Back is blue gray to olive gray, underside silver white to dirty white. DISTRIBUTION: In Virginia, this fish is restricted to the Clinch and Powell rivers, where it is rare or uncommon. HABITAT: The paddlefish occupies warm, medium-sized and large rivers and reservoirs. In rivers, it typically inhabits long, sluggish pools, backwaters, and oxbows. Paddlefish appear to prefer depths greater than 4 feet and seek deeper water in late fall and winter. LIFE HISTORY: An indiscriminate plankton feeder, the paddlefish usually feeds pelagically, straining plankton and aquatic insects across the extensive sieve formed by the numerous long gill rakers. Maturation occurs between years 7 and 12. Life expectancy is 15 years. Paddlefish gregariously spawn over clean gravel. THREATS: Threats to the habitat include siltation (soil and coal fines) and pollution. NOTES: The paddlefish is native to the Mississippi River basin and smaller Gulf of Mexico tributaries. It was extirpated from the Great Lakes basin by the late 1900s. Exploitation by the caviar industry is a recent threat in some parts of the US.

Abridged from "Paddlefish" by Noel M. Burkhead and Robert E. Jenkins, pp. 345–346 in *Virginia's Endangered Species*, coordinated by Karen Terwilliger, 1991 (Blacksburg, VA: The McDonald & Woodward Publishing Company).

Turquoise Shiner

Cyprinella monacha

THREATENED

DESCRIPTION: The turquoise shiner is elongate with a large caudal spot and blackened posterior dorsal fin membranes. Adults reach a length of 2¼ to 3½ inches. The dorsal fin origin lies slightly posterior to the pelvic fin origin; the snout is somewhat long, rounded or blunt; the mouth small, inferior, often sharply curved downward posteriorly. Adult females and non-nuptial males are tan, gray, or olive green dorsally; the rest of the head and body is bright silver. Nuptial males at peak have upper side of body iridescent turquoise or royal blue; irises silver or gold; all fins shiny turquoise with gold glints. DISTRIBUTION: The turquoise shiner is known from the North Fork of Holston River and in a short length of the Middle Fork of Holston River. It has not been seen in the North Fork above Saltville since 1954 and has not been seen at Saltville since 1888. However, it appears to be recolonizing the middle Washington County section of the North Fork. HABITAT: The turquoise shiner typically occurs in clear, cool and warm, large creeks to medium-sized rivers of moderate gradient. It favors moderate and swift currents over gravel, rubble or bedrock. LIFE HISTORY: Young and adults are benthic insectivores; immature midge and blackfly larvae comprise 90 percent of the diet. The spawning period is lengthy; nuptial adults may be seen from mid-May to mid-August. Females deposit eggs in crevices and cracks. It is likely that several clutches of eggs are produced in a spawning season by a each female. THREATS: Siltation, chemical pollution, and impoundment are the primary threats to this species. Overcollecting can threaten local subpopulations. NOTES: The turquoise shiner is endemic to the Tennessee drainage basin. It has been extirpated from Alabama and Georgia. At present populations are known to survive in only four stream systems. The species was listed as federally endangered on 9 September 1977. (This fish was formerly known as the spotfin chub, *Hybopsis monacha*.)

Abridged from "Turquoise Shiner" by Noel M. Burkhead and Robert E. Jenkins, pp. 346–347 in *Virginia's Endangered Species*, coordinated by Karen Terwilliger, 1991 (Blacksburg, VA: The McDonald & Woodward Publishing Company).

Steelcolor Shiner

Cyprinella whipplei

THREATENED

DESCRIPTION: The steelcolor shiner has a fairly deep body; the posterior or all dorsal fin membranes are dusky or black, and the dorsal fin is distinctly convex-margined in adult males. Adults are usually 2½ to 4½ inches long. The dorsal fin origin lies slightly posterior to pelvic fin origin. Juveniles and adult females are olive above, have sides with silver overlay, and white bellies. The back of nuptial male is steel blue with violet to rose sheens; sides and underside are silvery white, fins have pale yellow flush. **DISTRIBUTION:** In Virginia, the steelcolor shiner is known from only five sites in the Clinch River. **HABITAT:** The steelcolor shiner occupies warm, moderate to somewhat low gradient large creeks and medium-sized to large rivers. It is found in runs, pools, and backwaters over a variety of substrates. **LIFE HISTORY:** The steelcolor shiner is primarily a drift feeder, eating mostly terrestrial and aquatic insects. Spawning occurs from early June to mid-August, usually near riffles. Aggregated males defend small territories, while females generally hold nearby. Eggs are deposited above the bottom in crevices or furrows on logs and among tree roots. **THREATS:** Silt and other consequences of agricultural activities contribute to the general degradation of the Clinch River and threaten the habitat of the steelcolor shiner. **NOTES:** Impoundment has destroyed much of the habitat of the steelcolor shiner in the upper Tennessee drainage. The absence of this species from the Powell River in Virginia may be due to its preference for larger streams.

Abridged from "Steelcolor Shiner" by Noel M. Burkhead and Robert E. Jenkins, pp. 348–349 in *Virginia's Endangered Species*, coordinated by Karen Terwilliger, 1991 (Blacksburg, VA: The McDonald & Woodward Publishing Company).

Slender Chub

Erimystax cahni

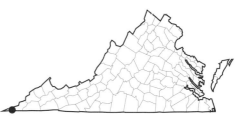

DESCRIPTION: The slender chub is elongate with a very long snout, barbeled mouth, and dusky lateral stripe often subdivided into a series of chevrons. Adults are usually 2 to 3¼ inches long. Dorsal fin origin is over or slightly anterior to pelvic fin origin. Head and body are yellowish tan to brown dorsally; lateral stripe and sometimes the area slightly above it are iridescent pale green. Lower side and belly are silvery white; when held at some angles, the side has violet iridescence. DISTRIBUTION: The slender chub was discovered in Virginia in 1979. It is known from only two sites in the Powell River, Lee County. Probably it was found in the Clinch River before the 1967 fish kill. HABITAT: The slender chub is a warm water riverine minnow. It is restricted to moderately to fast flowing shallow flats and shoals composed mostly of pea-sized gravel. LIFE HISTORY: The slender chub eats aquatic insect larvae and tiny snails and mussels. Spawning is believed to occur during late April and early May. THREATS: Siltation, dredging, pollution, water withdrawal, and impoundment are threats to the slender chub. The Powell River has been adversely impacted by deposition of coal fines from coal washing operations. Collecting is also a direct threat to the Virginia population. NOTES: The slender chub, known only from the upper Tennessee River drainage in Tennessee and Virginia, has one of the smallest ranges among North American riverine fishes. A main reason for the decline of the species is its strong predilection for fine-gravel shoals, a habitat often lost with impoundment or smothered by fine sediments. Of the imperiled main-channel fishes in the Clinch River, the slender chub is the rarest. It has not been seen since 1987, and some biologists fear it is at the brink of extinction. The species was federally listed as threatened on 9 September 1977. (It was formerly known as *Hybopsis cahni*.)

Abridged from "Slender Chub" by Noel M. Burkhead and Robert E. Jenkins, pp. 349–350 in *Virginia's Endangered Species*, coordinated by Karen Terwilliger, 1991 (Blacksburg, VA: The McDonald & Woodward Publishing Company).

Whitemouth Shiner

Notropis alborus

THREATENED

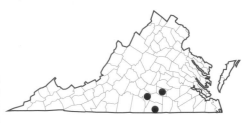

DESCRIPTION: The whitemouth shiner has a well rounded or blunt snout encircled by a dark band. The lateral stripe is prominently dark; a middorsal stripe is absent. Adults are 1½ to 2 inches long. Dorsal fin origin is over or slightly anterior to pelvic fin origin. Mouth small, nearly horizontal. The back is pale straw-colored, sometimes olive tinted, and lacks a middorsal stripe before the dorsal fin; lateral stripe black with steely blue violet iridescence. Lower side and underside are white with a hint of violet iridescence. **DISTRIBUTION:** The whitemouth shiner is known from the upper Nottoway River and from the Roanoke River drainage in Virginia. One of the few persisting populations in the state occupies Mines Creek, a tributary of Allens Creek of Gaston Reservoir. A population also survives in Aarons Creek in the Kerr Reservoir watershed. **HABITAT:** A species of the middle and lower Piedmont, the whitemouth shiner lives in warm, clear or somewhat turbid, small to medium-sized creeks. In Mines Creek, it is found in shallow small pools and in deep and shallow portions of long pools, in places floored with silt, sand, and bedrock. **LIFE HISTORY:** The natural history of the whitemouth shiner is poorly known. Its diet consists of microcrustaceans, mites, diatoms, and detritus. It likely spawns in late spring and early summer. **THREATS:** Impoundment, channelization, and possibly siltation and agricultural runoff are threats to the habitat of the whitemouth shiner. **NOTES:** The whitemouth shiner is also known from the Roanoke, Cape Fear, Peedee, and Santee river drainages of North Carolina. It appears to have vanished from the Roanoke Creek system in Virginia.

Abridged from "Whitemouth Shiner" by Noel M. Burkhead and Robert E. Jenkins, pp. 350–352 in *Virginia's Endangered Species*, coordinated by Karen Terwilliger, 1991 (Blacksburg, VA: The McDonald & Woodward Publishing Company).

Emerald Shiner

Notropis atherinoides

THREATENED

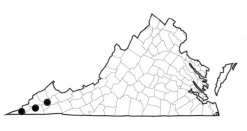

DESCRIPTION: The emerald shiner is slender, mostly pallid with dusky midlateral stripe and posteriorly positioned dorsal fin. Adults are usually 2¼ to 3 inches long. The back is translucent straw green to green, side with iridescent silver or green stripe, lower side and underside silver white. Fins are pale. Chromatic nuptial color is lacking. DISTRIBUTION: The species is known to occur in Virginia only in the Clinch and the Powell rivers. The population in the Powell River, however, may be extirpated. HABITAT: The emerald shiner lives in lakes and reservoirs and in pools and slows runs of river with low and moderate gradient. In Virginia, it occurs in small and medium-sized rivers. A pelagic minnow, this shiner often forms large schools in upper levels of water. LIFE HISTORY: In streams, the emerald shiner chiefly feeds in the middle and upper portions of the water column, on drifting terrestrial and aquatic insects. Most emerald shiners first spawn in their second year; longevity is greater than four years. THREATS: Specific threats to the emerald shiner are difficult to identify. It tolerates impoundments. Because the species is pelagic, substrate smothering silt may not be a particular problem, although it could result in high mortality of eggs. NOTES: The emerald shiner has one of the largest ranges of all North American minnows, inhabiting streams from the Gulf of Mexico northward into Canada. Why this fish, so common in many areas — in breeding schools they may number in the millions — fares so poorly in Virginia is not clear. One possible explanation is that the populations may be depressed by cumulative effects of low-level pollution.

Abridged from "Emerald Shiner" by Noel M. Burkhead and Robert E. Jenkins, pp. 353–354 in *Virginia's Endangered Species*, coordinated by Karen Terwilliger, 1991 (Blacksburg, VA: The McDonald & Woodward Publishing Company).

Tennessee Dace

Phoxinus tennesseensis

ENDANGERED

DESCRIPTION: The Tennessee dace is a fine-scaled minnow with two lateral stripes in adults. The upper stripe is tapered and subdivided posteriorly, the lower stripe partly or entirely disaligned at about midlength with a deflection toward the anal fin. Adults usually 1½ to 2¼ inches long. Juvenile is olive above, sides silver with one or two dusky or black lateral stripes, belly white. Nuptial male is dorsally tan or olive, usually with small dusky flecks; the stripe between the lateral black stripes flat to shiny silver or gray mustard; the lower side and belly diffuse to intense scarlet. Chin and breast sooty or black, lower fins lemon yellow, irises partly silver or yellow gold. Nuptial female is subdued, usually with little or no red or gold. **DISTRIBUTION:** Only two small headwater populations are known in Virginia. In the North Fork of Holston River system a population occupies the upper, Bland County, portion of Lick Creek. In the Middle Fork of Holston River the Tennessee dace is known from lower and middle Bear Creek and from the mainstem Middle Fork of Holston River near the mouth of Bear Creek. A population may also exist in Walker Creek, a tributary of the middle section of the Middle Fork of Holston River. **HABITAT:** The Tennessee dace is essentially restricted to small creeks containing cool and cold, typically clear water and having moderate gradient. It dwells in rocky, gravelly, and silty pools in wooded areas. **LIFE HISTORY:** Like other daces, the Tennessee dace probably eats living and decayed plant matter. They apparently do not spawn until two years of age and live three years. **THREATS:** Channelization, drying of streams, impoundment, siltation, and release of bait minnows into the streams it occupies are threats to the Tennessee dace. It may also be in danger from competition with introduced mountain redbelly dace, the presence of which probably stems from releases by bait fishermen.

NOTES: In spawning dress, the Tennessee dace is one of the most brilliantly and ornately colored fishes in Virginia. Only two small populations of the Tennessee dace are definitely known in Virginia. Both are isolated in separate stream systems, hence they cannot contribute to each other's survival by natural dispersal. The habitat of one population is degraded; that of the other is threatened with degradation.

Abridged from "Tennessee Dace" by Noel M. Burkhead and Robert E. Jenkins, pp. 360–362 in *Virginia's Endangered Species*, coordinated by Karen Terwilliger, 1991 (Blacksburg, VA: The McDonald & Woodward Publishing Company).

Yellowfin Madtom

Noturus flavipinnis

<small>THREATENED</small>

DESCRIPTION: The yellowfin madtom is a saddled, blotch-finned madtom catfish; adults are usually 3 to 4½ inches long. Body ground color is pale yellow brown to yellow gray, with slight pink cast midlaterally. All fins are very pale yellow along base, except the pelvic fin, which is translucent white with pinkish cast. **DISTRIBUTION:** The single extant Virginia population inhabits lower and middle Copper Creek in Scott and Russell counties. This madtom was known from the North Fork of Holston River just above Saltville. It may occur in the Clinch River near Copper Creek, and the population in the Powell River, Tennessee, may extend into Virginia. Yellowfin madtoms are generally rare, but populations are difficult to inventory because of their secretive, nocturnal habits. **HABITAT:** The yellowfin madtom occupies unpolluted medium-sized and large creeks to small rivers with moderate to gentle gradient. It is almost always found in calm water, usually in slow pools. During daylight it seeks cover under sticks, logs, leaf litter, undercut banks, rocks, and trash. It is found in open, clean gravel, and rubble areas only at night. **LIFE HISTORY:** The yellowfin madtom eats a wide variety of immature benthic insects. It lives three, possibly four years, with most individuals maturing in two years. Spawning occurs from mid-May to July. Eggs are deposited in cavities beneath flat rocks in pools at depths less that 3 feet. **THREATS:** The habitat of the yellowfin madtom is degraded by siltation, agricultural pollution, and impoundment. **NOTES:** The yellowfin madtom was thought to be extinct when it was first described in 1969 from specimens collected between 1884 and 1893. The species was federally listed as threatened in 1977. As part of the recovery plan, the US Fish and Wildlife Service and University of Tennessee have been transplanting this madtom within Tennessee. They will attempt to reestablish a population in the North Fork of Holston River.

Abridged from "Yellowfin Madtom" by Noel M. Burkhead and Robert E. Jenkins, pp. 365–366 in *Virginia's Endangered Species*, coordinated by Karen Terwilliger, 1991 (Blacksburg, VA: The McDonald & Woodward Publishing Company).

Orangefin Madtom

Noturus gilberti

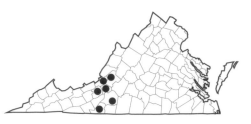

THREATENED

DESCRIPTION: The orangefin madtom is a catfish with an unpatterned
body and somewhat triangular pale area
on the outer portion of the upper cau-
dal fin lobe. Adults are usually 2½ to 3¼ inches long. The upper half of head and body is
pale gray to medium brown, often with slight yellow olive cast. Underside is white, with
slight pink cast in some individuals; pale areas of fins are colorless or with slight to moderate yellow wash, never bright orange, except when fins are folded. DISTRIBUTION:
The orangefin madtom is native to the upper Roanoke River drainage. Only five isolated
indigenous populations exist, the clearly viable ones inhabiting the Roanoke River from
Salem upstream through the South Fork of Roanoke River into lower Bottom and Goose
Creeks; lower Big Chestnut Creek and nearby Pigg River; the Dan River from its gorge
in the Blue Ridge downstream into North Carolina; and the South Mayo River and North
Fork of South Mayo River within or just above Stuart. The Craig Creek population, in
the James River drainage, was almost certainly transplanted from the Roanoke River drainage. HABITAT: Orangefin madtoms occupy a narrow range of habitat and occur in medium-sized, moderate gradient, montane and upper Piedmont streams. The largest
populations occupy generally clear waters. The orangefin madtom shows an affinity for
moderate and strong runs and riffles, where it dwells in cavities amidst the rubble and
boulders. LIFE HISTORY: The orangefin madtom feeds almost entirely on immature
aquatic insects, particularly mayflies, caddisflies, and midges. At least some feeding occurs at night. Spawning apparently occurs from late April through May. Breeding sites
are probably beneath loose
rubble. The species is short-
lived. THREATS: Siltation is
a pervasive problem throughout the species' range. Bait-
seining is a direct threat. The
apparently low reproductive
potential of the orangefin
madtom makes it particularly
vulnerable to disturbance.
NOTES: The orangefin mad-
tom is an evolutionary and
zoogeographic relict that has
survived a long period of geologic time in the upper Roanoke drainage.

Abridged from "Orangefin Madtom" by Noel M. Burkhead and Robert E. Jenkins, pp. 367–369 in *Virginia's Endangered Species*,
coordinated by Karen Terwilliger, 1991 (Blacksburg, VA: The McDonald & Woodward Publishing Company).

Blackbanded Sunfish

Enneacanthus chaetodon

ENDANGERED

DESCRIPTION: The blackbanded sun- fish is a small sunfish with bold black bars on the body, nonspotted fins, and adult length of 1¼ to 2½ inches. Body ground color is mostly opalescent white with brassy sheens, back medium olive brown except for bars, olive brown variegations on upper half of body. Irises are orange copper, as are pelvic spine and leading pelvic ray. DISTRIBUTION: Only three sharply localized populations are known in Virginia, where it occurs in both the Nottoway and Blackwater river systems of the Chowan River drainage. HABITAT: The blackbanded sunfish is a denizen of ponds, swamps, and streams. An important habitat requisite seems to be beds of submerged vegetation, which probably provide shelter and spawning sites and which harbor its prey. Typically this species is associated with very acidic waters of pH 4.1 to 6.6. LIFE HISTORY: The diet of the blackbanded sunfish consists of an array of small invertebrates associated with aquatic macrophytes. Nests vary from small dish-shaped depressions formed in the substrate beneath vegetation to hollows made in plants. Lon- gevity is three to four years. THREATS: Drying of ponds and swamps and contamination of water by herbicides and pesticides are potential threats to the habitat of the blackbanded sunfish. Taking of individuals for aquaria could place populations in jeopardy. NOTES: The blackbanded sunfish has been sold recently as "banded sunfish" in pet stores in the Norfolk-Virginia Beach area. (The true banded sunfish, *Enneacanthus obesus*, is not endangered.) There are re- ports of the blackbanded sun- fish having been exported to Taiwan to be raised for the aquarium industry. Black- banded sunfishes confiscated from pet stores should not be released into the wild, because they may have originated from a geographically distant population and mixing could cause genetic disharmony if they were to interbreed with indigenous fishes.

Abridged from "Blackbanded Sunfish" by Noel M. Burkhead and Robert E. Jenkins, pp. 375–377 in *Virginia's Endangered Species*, coordinated by Karen Terwilliger, 1991 (Blacksburg, VA: The McDonald & Woodward Publishing Company).

Western Sand Darter

Ammocrypta clara

THREATENED

DESCRIPTION: The western sand darter is very slender and pallid with a long, sharp snout. Adults measure 1½ to 2¼ inches long. The opercular spine is well developed; caudal fin slightly emarginate; lateral line complete. Body is very translucent, grayish; cheek and opercle with subtle blue green iridescence; blue green iridescence less evident along midlateral body. **DISTRIBUTION:** The Virginia records in the Clinch and Powell rivers mark the upper limit of the western sand darter in the Tennessee River drainage. The sand darter was first found in Virginia in 1979. It is likely that it had been extirpated from the Clinch River by the 1967 fish kill and has become reestablished from populations in Tennessee or from those at mouths of tributaries of the Clinch River in Virginia. It is seemingly rare at all three sites of record in Virginia. **HABITAT:** The western sand darter inhabits medium and large rivers with warm waters and low or moderate gradients. It typically occupies sand substrate in moderate current. **LIFE HISTORY:** The main dietary components of the western sand darter are larvae of aquatic insects. **THREATS:** Siltation, chemical spills, and agricultural practices are threats to the habitat of the western sand darter. **NOTES:** The reappearance of the western sand darter in Virginia is an example of faunal recovery after a catastrophic chemical spill. This darter is the only member of its genus in Virginia.

Abridged from "Western Sand Darter" by Noel M. Burkhead and Robert E. Jenkins, pp. 377–378 in *Virginia's Endangered Species*, coordinated by Karen Terwilliger, 1991 (Blacksburg, VA: The McDonald & Woodward Publishing Company).

Sharphead Darter

Etheostoma acuticeps

ENDANGERED

DESCRIPTION: The sharphead darter is obliquely barred, plain-finned, and sharp-snouted; adults are usually 1¼ to 2½ inches long. Adults other than nuptial males have head and body mostly 'straw olive to brown olive. Nuptial male has dark olive back, tan underside, sides occasionally with turquoise tinge, and breast blue green. The first dorsal fin is distally with suffuse turquoise, the remainder of the fin is dark olive. Other fins are bright turquoise. DISTRIBUTION: The single population in Virginia inhabits the South Fork of Holston River. HABITAT: The sharphead darter inhabits large creeks to medium-sized rivers that are cool or warm, usually clear or slightly turbid, and have moderate gradients. It usually occupies moderate and swift runs and riffles of rubble and boulder and often occurs among lush growth of riverweed. It has an affinity for swift chute-like water in the South Fork of Holston River. LIFE HISTORY: Mayfly, midge, and blackfly larvae dominate the diet. Both sexes mature as one-year-olds, and some may live for three years. Sharphead darters in breeding condition have been reported between late July and mid-August. Indirect evidence suggests this darter is an egg-burier. THREATS: Siltation and other forms of pollution are major threats to the sharphead darter in the South Fork of Holston River. NOTES: The sharphead darter is endemic to the upper Tennessee River drainage. It is known to inhabit four streams in North Carolina, Tennessee, and Virginia. The many fishermen of the lower South Fork of Holston River would be astonished by the sight of the strikingly turquoise-hued nuptial male sharphead darter.

Abridged from "Sharphead Darter" by Noel M. Burkhead and Robert E. Jenkins, pp. 378–380 in *Virginia's Endangered Species*, coordinated by Karen Terwilliger, 1991 (Blacksburg, VA: The McDonald & Woodward Publishing Company).

Greenfin Darter

Etheostoma chlorobranchium

THREATENED

DESCRIPTION: The greenfin darter is lined, checkered, or faintly barred, the second dorsal, caudal, and anal fins margined with black and usually bright green next to the margin. Adults measure 1¾ to 2¾ inches long. Juveniles, adult females, and nonbreeding adult males mostly drab olive, but with contrasting fin margins and, in adults, red spots on body. Nuptial males typically green to blue green on head, body, and much of fins, and blue on breast. DISTRIBUTION: In Virginia, the greenfin darter is known only from the lower part of Whitetop Laurel Creek, a tributary to Laurel Creek of the South Fork of Holston River system. HABITAT: The greenfin darter inhabits small and large streams with clear, cold or warm waters and moderate and high gradients. In Virginia, specimens have been captured from riffles of boulders and loose rubble. LIFE HISTORY: The life history of this species is largely unknown. THREATS: No major anthropogenic threats are known. The watershed of Whitetop Laurel Creek is protected as part of the Mount Rogers National Recreation Area. NOTES: The greenfin darter is endemic to the portion of the upper Tennessee drainage in the Blue Ridge of Georgia, North Carolina, Tennessee, and Virginia. The Whitetop Laurel Creek population is at the northern terminus of its range, and greenfin darters are extremely rare there. Fewer than ten specimens are known from Virginia.

Abridged from "Greenfin Darter" by Noel M. Burkhead and Robert E. Jenkins, pp. 381–382 in *Virginia's Endangered Species*, coordinated by Karen Terwilliger, 1991 (Blacksburg, VA: The McDonald & Woodward Publishing Company).

Carolina Darter

Etheostoma collis

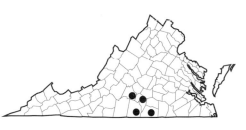

DESCRIPTION: The Carolina darter is laterally blotched or flecked with a short pale lateral line high on the body, and usually one anal spine. Adults are 1 to 1½ inches long. Spawning males are moderately dark with lateral marks that are black olive or brown olive. The back is pale olive with medium to black live marks; lower head and pectoral base has green iridescence; irises partly coppery. Females have lighter olive lateral marks; and back is straw-colored with medium olive marks. Belly has light gold wash. DISTRIBUTION: The Carolina darter occurs as localized populations in the Roanoke River drainage from tributaries of the upper Gaston Reservoir to the lower part of Falling River. It is generally rare, but common to abundant in Mines Creek, Mecklenburg County. HABITAT: The tiny Carolina darter occupies small creeks and rivulets in wooded and deforested areas. It lives in open and stick-littered portions of pools and very slow runs, usually on sand, gravel, and detritus substrates. It probably also dwells among vegetation. Spawning sites are gravel, sticks, leaves, and aquatic moss. LIFE HISTORY: The Carolina darter feeds on microcrustaceans and mayfly and dipteran larvae. Males solicit females with fin erections and contact with head and fins. Usually a female is mounted by one male; frequently one or two other males join the pair at the initiation of spawning. Some of the sneaker or satellite males may be a pale shade similar to that of females. THREATS: Chemical runoff from agricultural land may be a factor limiting this species. In addition, this sight-feeding fish may be adversely affected by moderate and high levels of turbidity caused by excessive amounts of silt in the waters. NOTES: The Carolina darter may have some tolerance to impoundment as it belongs to a group of darters adapted to lowland conditions. Such darters commonly live in waterways that contain much sediment and have little or no current. The Carolina darter is adept at perching on and attaching eggs to sticks above the bottom.

Abridged from "Carolina Darter" by Noel M. Burkhead and Robert E. Jenkins, pp. 383–384 in *Virginia's Endangered Species*, coordinated by Karen Terwilliger, 1991 (Blacksburg, VA: The McDonald & Woodward Publishing Company).

Tippecanoe Darter

Etheostoma tippecanoe

DESCRIPTION: The Tippecanoe darter is tiny, speckled or barred, usually with a prominent saddle over anterior body. Adult length ¾ to 1¼ inches. Female is subdued, generally a mottled olive brown, sometimes with a faint gold cast. Breeding male with bars and breast blue to black, elsewhere body side olive yellow to moderately orange, belly sometimes has pale blue wash; opercle and gular area gold to deep red orange; lips yellow orange. Some or all fins partly yellow olive, gold, or red orange; caudal fin base has two yellow gold patches. In the spawning male, the orange intensifies on the body, obscuring most bars. **DISTRIBUTION:** In Virginia, the Tippecanoe darter is known only from four sites on the Clinch River and from the lowermost part of Copper Creek. The absence of this species from the Powell River probably is due to extirpation. **HABITAT:** Tippecanoe darters inhabit medium-sized rivers with moderate gradients and water that is warm and usually clear. Adults occupy shallow and deep, moderate and swift, runs and riffles, associating particularly with large areas of unsilted pea-sized gravel. **LIFE HISTORY:** The diet of this darter consists mainly of mayfly, caddisfly, and midge larvae. The species exhibits major annual fluctuations in population size and may be an annual species. Breeding probably occurs in July and extends into August. Spawning occurs in runs and riffles of gravel or sand-gravel substrates. Males are territorial and females are egg-buriers. **THREATS:** Siltation of the Clinch River is a chronic limiting factor to the Tippecanoe darter, as may be contaminants in runoff from agricultural areas. **NOTES:** The Tippecanoe darter was discovered in the Clinch River one month before its extirpation due to a vast toxic spill at Carbo in June 1967. Recolonization of the Clinch River has occurred — probably from downstream reaches of the river in Tennessee, but the Tippecanoe darter has not been found in Copper Creek since 1967.

Abridged from "Tippecanoe Darter" by Noel M. Burkhead and Robert E. Jenkins, pp. 387–388 in *Virginia's Endangered Species*, coordinated by Karen Terwilliger, 1991 (Blacksburg, VA: The McDonald & Woodward Publishing Company).

Variegate Darter

Etheostoma variatum

ENDANGERED

DESCRIPTION: The variegate darter is laterally speckled or barred with four saddles; adults are 2 to 3 inches long. Body form is moderate to somewhat robust. Adult male has olive back, light green belly, white breast, and lateral bars green to blue green with interspaces yellow with orange or red spots. A bright orange red or red stripe appears on anterior flank. First dorsal fin is orange red submarginally, below which is blue black, basally red brown. Second dorsal fin has dusky blue margin, orange red speckling, and mostly dusky olive green; caudal margin and base dusky blue black, centrally orange green with red spots. Anal and pelvic fins are mostly dusky green; pectoral fin olive has red or brown spots. In the nuptial male, blue green, orange, and red are intensified; in the female chromatic colors are subdued or largely olive. DISTRIBUTION: The variegate darter occupies only in the Big Sandy River drainage; the four records are from Levisa Fork, Long Branch, Dismal Creek, and Knox Creek. HABITAT: Variegate darters occupy warm small and large streams and tend to localize in riffles of gravel, rubble, boulder, and some sand, where there is little siltation. LIFE HISTORY: The diet of the variegate darter includes aquatic insect larvae and water mites. The species first spawns at age two and lives three years or so. THREATS: Siltation and acidification of streams from coal mine wastes are forms of habitat degradation in the Big Sandy River drainage of Virginia and are limiting factors for the variegate darter. NOTES: The variegate darter is widespread in the upper and middle Ohio River basin, exclusive of the New River. It has apparently been extirpated from much of the Big Sandy River drainage in Virginia. Even with strict erosion control measures, the Big Sandy River system would require decades to flush itself. The survival of populations of this colorful darter in Virginia is tenuous. It is one of the rarest fishes in the state.

Abridged from "Variegate Darter" by Noel M. Burkhead and Robert E. Jenkins, pp. 388–389 in *Virginia's Endangered Species*, coordinated by Karen Terwilliger, 1991 (Blacksburg, VA: The McDonald & Woodward Publishing Company).

Duskytail Darter

Etheostoma sp.

DESCRIPTION: The duskytail darter is distinguished from the common fantail darter, *Etheostoma flabellare,* by narrower bars, a pectoral fin that is pigmented only distally, a head often with large freckles, first saddle often yoke-like. Adults measure 1 to 1¾ inches. In both sexes the back and side ground color is straw to brown olive or gray olive, sometimes with a pale yellow wash; saddles, bars, and top of head medium to dark olive; undersides are dingy white to pale gray. In nuptial male, irises are golden, and anal fin is creamy white with sharply contrasting black margin. **DISTRIBUTION:** One relict population is known in Virginia in Copper Creek, where it inhabits the lower main channel. **HABITAT:** The duskytail darter occupies upland and montane medium to large creeks and medium-sized rivers that are of moderate gradient, warm, and usually clear. The species predominates in pools with gravel, rubble, and/or boulder bottoms; it often is found among detritus and in slightly silted areas. **LIFE HISTORY:** The duskytail is a benthic invertivore. Smaller individuals feed mostly on crustaceans and midge larvae; larger ones consume midge, mayfly, and caddisfly larvae. Most spawning seemingly takes place in May. The duskytail is an egg-clusterer, spawning during a lengthy period of inversion under a stone. The male cleans and aggressively protects the nest cavity. **THREATS:** Silt, runoff from agricultural activities, and impoundment are threats to the duskytail darter in Copper Creek. Competition with the closely related fantail darter is a possible threat. **NOTES:** The duskytail darter is endemic to the upper Tennessee and middle Cumberland river drainages. Six relict populations are known: the one in Virginia and five in Tennessee. Two of the Tennessee populations have been extirpated.

Duskytail and fantail darters have complementary distributions in Copper Creek, the fantail occupying the middle and upper portions and major tributaries. In zones of overlap, the fantail is the rarer species.

Abridged from "Duskytail Darter" by Noel M. Burkhead and Robert E. Jenkins, pp. 389–391 in *Virginia's Endangered Species,* coordinated by Karen Terwilliger, 1991 (Blacksburg, VA: The McDonald & Woodward Publishing Company).

Longhead Darter

Percina macrocephala

THREATENED

DESCRIPTION: The longhead darter has a fairly long, subconic snout, and prominent broadly conjoined lateral blotches. Adults are usually 2½ to 3½ inches long. The back and upper side are olive to brown with markings brown to black. A narrow yellow stripe occurs just above the lateral blotches. The lower side and venter are yellow to white. The pale band of the first dorsal fin lacks bright orange; other fins are commonly pale yellow to olive. DISTRIBUTION: In Virginia, the longhead darter is known from two tributaries of the Clinch River — Copper Creek and Little River — and in the North and Middle Forks of Holston River. Since 1970 the longhead darter has been verified only in lower Copper Creek and the North Fork of Holston River above Saltville. HABITAT: The longhead darter inhabits usually clear, medium-sized streams of moderate gradient. In much of its range it occupies well flowing pools, runs, and riffles with substrates ranging from bedrock to those supporting vegetation; the apparent favored substrate is clean and rocky. LIFE HISTORY: Foods include mayfly larvae and crayfishes. Lifespan is three to four years. Spawning probably occurs in the spring. Threats. Little is known about threats to the longhead darter in Virginia. Adult longhead darters may be fairly tolerant of silt, but eggs and larvae might be vulnerable to smothering by silt. Impoundment would almost certainly cause extirpation of affected populations. NOTES:

The longhead darter has a spotty distribution in the Ohio River basin. It was taken in the Middle Fork of Holston River only during 1888 and 1937. It may have occupied the South Fork of Holston before impoundment of that river in Tennessee. The North Fork of Holston population should be increasing following recovery of the river from chemical contamination, but there is no clear evidence of population growth.

Abridged from "Longhead Darter" by Noel M. Burkhead and Robert E. Jenkins, pp. 394–395 in *Virginia's Endangered Species*, coordinated by Karen Terwilliger, 1991 (Blacksburg, VA: The McDonald & Woodward Publishing Company).

Roanoke Logperch

Percina rex

ENDANGERED

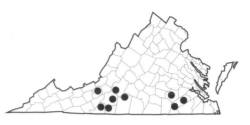

DESCRIPTION: The Roanoke logperch has round or vertically elongate lateral blotches, a back much scrawled, most fins strongly patterned, and snout moderate or long, conical or piglike. Adults measure 3¼ to 4½ inches. Head and upper half of body is straw olive in juveniles and adults, lower body straw to whitish, back with brown scrawl lines, sides with brown black bars; gold, green, or blue iridescence occurs on side of head, in the pre-pectoral area, and in lateral blotches. The first dorsal fin has a submarginal yellow or orange band. Nuptial male has heightened melanistic and chromatic pigmentation. DISTRIBUTION: The Roanoke logperch is endemic to the Roanoke and Chowan River drainages of Virginia. In the Ridge and Valley, it is continuously distributed in the upper Roanoke River and lower North Fork and South Fork of Roanoke River and is known from lower Mason and Tinker creeks. It sparsely inhabits the Pigg River and the extreme lower reach of Big Chestnut Creek. The upper Smith River and lower Town Creek of the upper Dan River system have small populations. In the Chowan River system it is known only from the Nottoway River system, in the Nottoway River, Stony Creek, and Sappony Creek. HABITAT: The Roanoke logperch inhabits medium-sized streams that are warm, usually clear, and have moderate to low gradient. Young and small juveniles usually occupy slow runs pools, most frequently sandy areas. During warmer months, adults typically dwell on gravel and rubble in riffles, runs, and pools. LIFE HISTORY: The Roanoke logperch feeds benthically, chiefly on immature insects. Young and juveniles prey heavily on midge larvae, adults mainly on caddisfly and midge larvae. Over-turning gravel and small rocks with the snout and eating the exposed prey is a primary foraging tactic. Males mature in two years, female three. Longevity may be up to 6 1/2 years. Spawning occurs from mid-April to early May. THREATS: Channelization, siltation, various forms of chronic pollution, catastrophic chemical spills, impoundment and dewatering are all threats to the Roanoke logperch. NOTES: The species was federally listed as endangered in 1989.

Abridged from "Roanoke Logperch" by Noel M. Burkhead and Robert E. Jenkins, pp. 395–397 in *Virginia's Endangered Species*, coordinated by Karen Terwilliger, 1991 (Blacksburg, VA: The McDonald & Woodward Publishing Company).

Amphibians and Reptiles

The ranges of no fewer than 134 species of amphibians and reptiles (collectively known as the herpetofauna) extend into or are contained entirely within the boundaries of Virginia. Several species reach the northern limits of their distributions in southeastern or southwestern Virginia, others reach their southern limits in northern Virginia, and several reach their eastern limits in western Virginia. Still others occur statewide. Two species are endemic to the state.

The composition of the herpetofauna appears to have been stable over the past 10,000 years. None of the species found in the late Pleistocene of Virginia has become extinct. However, distributions of species have undoubtedly shifted during this time. Isolated populations occurring some distances away from the main body of the current range (for example, the tiger salamander in Augusta County) provide circumstantial evidence for the shifting ranges.

Although it is difficult to assess changes in herptile populations during prehistoric times, there has been a dramatic decline in the number of viable populations since the onset of European colonization in Virginia nearly 400 years ago. The clearing of land for timber and agriculture was rapid, and by the middle of the 19th century, most of Virginia east of the mountains had been cleared of its forest. Loss of soil, siltation, construction of impoundments, and drainage of wetlands changed the environment of much of Virginia. No doubt many local populations of amphibians and reptiles have been extirpated. The increased urbanization, industrialization, and mechanization of agriculture following the world wars further added to the decline of herptile populations. Thus, the stage was set long ago that created our present concern about the future of these animals.

Identification of amphibians requires attention to patterns and morphology. Anurans, the frogs and toads, are superficially similar in appearance and possess many of the same features. Salamanders are considerably more difficult to distinguish. They differ from lizards, with which they are often confused, by having a generally smooth, moist skin, and toes without claws. If the pattern and coloration do not distinguish a salamander, the number of costal grooves and the number of costal folds between adpressed limbs should. Positive identification of some individuals may require taking the animal to someone who has considerable experience with the group.

Reptiles differ from amphibians in possessing a relatively hard skin-covering of scales or scutes. The arrangements of the various scutes, or plates, on the shells of sea turtles are especially useful for identification. This, combined with color patterns, should allow one to identify most specimens of sea turtles. The freshwater turtles covered in this chapter occupy non-overlapping ranges in Virginia; along with color and pattern characteristics, this fact should allow their correct identification. Recognition of most snakes requires knowledge of the differences in patterns and scales. The canebrake rattlesnake, for example, is easily identified by its black chevrons, rattle on the end of the tail and facial heat sensing pit. The only lizard included in this chapter is the eastern glass lizard. It is easily distinguished from most other lizards because it lacks legs. The hard, square-shaped scales, a moveable eyelid, and an ear opening distinguish this legless lizard from snakes.

Amphibians and reptiles represent two major phases in the evolution of vertebrates. Amphibians are vertebrates that are tied to water for reproduction. Reptiles were liberated from that constraint by the appearance of the amniotic, or shelled, egg. The development of the shelled egg enabled reptiles to colonize many terrestrial habitats unavailable to amphibians.

The anurans exemplify the general pattern of amphibian reproduction. Males and females shed spermatozoa and eggs simultaneously, and fertilization takes place externally, in the water. Eggs are enclosed within a gelatinous envelope and either adhere to one another in subsurface clumps and surface masses or sink to the bottom singly and in small clusters. There is no parental care exhibited by Virginia's frogs, although parents in several species around the world do extend such care. The eggs develop into aquatic, herbivorous larvae. This aquatic stage persists for a variable length of time, depending upon the species, from 14 days to two years. Length of the larval period, size at metamorphosis, and number of surviving individuals depend on densities of larvae and their predators, availability of food, and time at which ponds dry. A frog's experiences during the larval period may influence some aspects (for example, survivorship) of its life after metamorphosis, as does the larval period in salamanders. Metamorphosis consists of a dramatic change from a gill-breathing body form adapted for swimming to a lung-breathing body form adapted for jumping on land. Transformation takes up to several days, and the result is a small version of the adult. Growth of individuals from metamorphosis to reproductive maturity is not well studied but probably takes eight to twenty months depending on the season when transformations took place and rate of growth. Adult males vocalize, producing species-specific advertisement calls. Each species exhibits its own pattern of mate selection and mating behavior. The location and nature of the breeding site — ephemeral pool, permanent pond, or stream — vary according to the habitat requirements of the species for egg laying and larval growth. The contrasting stages in the life of a single frog constitute a complex life history strategy. Amphibians with such strategies are exposed to the hazards of the aquatic environment during the larval stages and the hazards of the terrestrial environment as adults.

Salamanders show a broad diversity of reproductive strategies, but many exhibit a complex life history similar to that of anurans. In general, the larval stage lasts from one to several years, and metamorphosis is not as dramatic. Mating behavior in salamanders is based on olfactory and visual cues rather than vocalizations. Males produce spermatophores, packets of spermatozoa, which are taken by females from the gelatinous stalks which males deposit on the forest floor or pool substrate. Fertilization is internal. Females of those species which reproduce in water, such as *Ambystoma*, lay eggs in clusters or masses which are usually affixed to a grass stem or twig. Several species of salamanders, notably species in the genus *Plethodon*, do not reproduce in water. These salamanders live in the litter

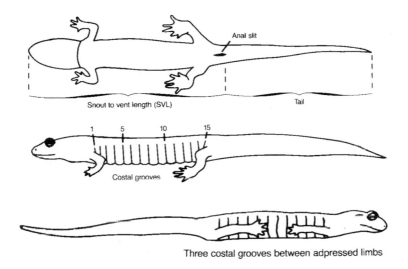

Figure 6. Generalized salamander with selected morphological characters.

and humus layers of the deciduous forest floor. Females of these species produce fewer eggs than do aquatic breeders, attach them to surfaces (such as under a rock), and guard them until hatching. All larval development takes place within the egg, and the hatchling is a miniature adult.

Anurans disperse over greater distances than salamanders. Most species of frogs can be found occasionally in unsuitable habitats far from water. Most salamanders, however, seldom disperse more than several yards during their entire lives. This is especially true for the genus *Plethodon*. Thus, colonization of newly created habitats could be relatively rapid for frogs and toads but very slow for salamanders. Because plethodontid salamanders do not disperse widely, fragmentation of their habitats has relatively more serious consequences for the long-term viability of isolated populations.

Amphibians are particularly vulnerable to stresses on their habitats because of their complex life histories. At some time in the life of all frogs and most salamanders each individual is subjected to the stresses found in both aquatic and terrestrial habitats.

Life histories of reptiles are substantially different from those of amphibians. All turtles lay shelled eggs. Most species mate in water, and males of all species possess intromittent organs allowing for internal fertilization. During the egg-laying season, females leave the aquatic habitat and seek suitable nesting sites on land. For sea turtles, this is usually a sandy beach. For freshwater turtles, it can be almost any area providing that the substrate can be dug. Flask-shaped nests are dug with the hind limbs, and a number of eggs — from one to several hundred —

are deposited. The female covers the nest and returns to water. There is no parental care. Some species do not lay eggs in nests but prefer to deposit them in grass clumps or under the edges of logs. The nesting season is a particularly vulnerable time for females due to predation by raccoons. Predation on eggs is usually very high and can sometimes result in zero survivorship. Eggs that survive predation incubate for two to several months The eggs or hatchlings of numerous species overwinter in the nest and emerge the following spring. Sea turtle hatchlings emerge from the nest *en masse* at the same time, and the hatchlings of several freshwater species (for example, the bog turtle) emerge from the nest in the fall rather than the following spring. Growth of hatchlings is rapid until reproductive maturity. In freshwater species, this usually takes three to five years for males and four to ten years for females. Although there is considerable speculation on this subject, age at maturity for sea turtles is unknown. Adult turtles are long-lived, to at least 20 years for many individuals; some have been recorded to have lived as long as 100 years. Such long-lived vertebrates may accumulate stresses, such as the storage of pesticides and other toxic chemicals in body tissues, that over time may reduce reproductive output and the viability of populations.

Lizards and snakes exhibit two reproductive strategies, egg-laying, or oviparity, and live-bearing, or viviparity. All lizard species in Virginia are oviparous. Many snakes lay shelled eggs, but the canebrake rattlesnake bears live offspring. New-

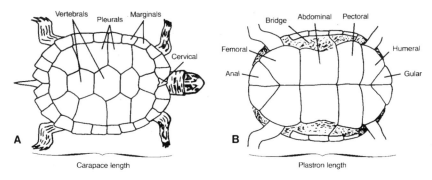

Figure 7. A generalized turtle indicating epidermal scutes of the (A) carapace and (B) plastron.

born rattlesnakes have functional venom glands. Growth rates, age at maturity, and length of life is not known for most snakes because of the difficulty of studying these secretive animals in the field.

The ability to colonize nearby or newly created habitats varies considerably in reptiles. Individuals of most freshwater turtles disperse over long distances — up to a few miles — at various times during the active season. Most, however, remain in the aquatic habitat or area in which they were produced. Snakes are known to disperse as juveniles and adults, but rates of dispersal and distances traveled are unknown for most species. The dispersal period represents a time of great stress, and evidence of the high mortality rates for dispersing reptiles can be found on Virginia's highways.

Fourteen species of herptiles that occur in Virginia are currently protected by federal and state endangered species laws. These include the five sea turtles that enter the Chesapeake Bay and Virginia's offshore waters, three freshwater turtles, one snake, one lizard, one frog, and three salamanders. The sea turtles and the Shenandoah salamander fall under the joint jurisdiction of the United States Fish and Wildlife Service and the Virginia Department of Game and Inland Fisheries. The others fall under the regulatory jurisdiction of the Virginia Department of Game and Inland Fisheries as state listed species.

Threats to the long-term viability of Virginia's rare amphibians and reptiles occur at the individual, population, and community levels. How successful our efforts will be at conserving Virginia's herptiles depends on how well we can understand all of the threats to each species concerned.

Some species of amphibians and reptiles are threatened with overcollecting for commercial purposes. Removal of an individual from its population for the pet trade has the same effect as that animal's being lost to predators. It can no longer contribute its genes to the following generation. The bog turtle and the canebrake rattlesnake currently face the threat of overcollecting in Virginia. Overcollecting of some species for scientific purposes has probably occurred in the past, but this is far better controlled than the commercial collecting has been. There is no evidence that scientific collecting has contributed to the decline in any species discussed in this chapter. Loss of individuals also occurs when humans kill them in their natural habitat and when migrating animals are killed on highways.

The primary threat to individuals and ultimately to populations of rare amphibians and reptiles is loss of habitat. This is the cause of danger to species most often invoked to support recommendations for endangered, threatened, or special concern status. Habitat loss has accelerated over the past several decades. Filling or draining wetlands and paving the natural landscape with concrete or asphalt as in urbanization usually kill outright most of the amphibians and reptiles on a site. Most of these animals cannot fly or run away from the heavy machinery as can birds and many mammals. Most suffer slow death from suffocation, fire, or other trauma. Thus, habitat loss is a direct cause of the decline in the numbers of individuals and populations of these vertebrates in Virginia.

Various forms of habitat alteration ultimately have effects similar to outright habitat loss. Clearcutting and mechanized selective cutting of timber alter the thermal regime of the forest floor, usually because more sunlight penetrates to the land surface where it heats and dries the microhabitat. Many amphibians are found in moist patches within the forest except in times of high moisture levels. Deforestation of any kind reduces the number of moist areas and increases the distances between them. Thus, subgroups of amphibians became scattered and isolated in what may appear to our eyes as a natural habitat. Drying of the forest floor is particularly deleterious to terrestrial salamanders which require moist skin for respiration.

Clearcutting with concomitant burning and disking of the soil is especially harmful to terrestrial amphibians and reptiles. This practice has been shown to destroy all the subterranean animals and render the habitat essentially sterile.

Fragmentation of habitat, by producing a patchwork of deforested areas and barriers such as roads and dams, creates several small populations from what was formerly a large one. Small populations are subject to a variety of phenomena, such as disease, environmental disasters, and inbreeding, that render them relatively more vulnerable to extinction than larger populations. It is not possible to pinpoint where fragmentation of the habitat has caused local population extinctions in Virginia because none of the herpetofauna has yet been studied from this perspective. However, terrestrial salamanders are most likely to be seriously affected.

Perturbations of aquatic systems directly or indirectly affect all inhabitants. Siltation resulting from poor land use practices and deforestation has reduced the amount of suitable habitat for aquatic species in two ways. The silt has caused populations of freshwater mussels, which are prey for several turtles, to decline. Silt covers rocky habitats and reduces prey populations and may cause respiratory stress and ultimately death by lowering oxygen levels. A variety of pesticides, herbicides, and other toxic pollutants harms the entire aquatic community. Developing embryos are the most sensitive stage of life and are killed by some chemicals. Invertebrates, often more susceptible to chemical pollutants than vertebrates, are prey and declines in prey populations indirectly harm members of the herpetofaunal community.

Entanglement, ingestion, and other complications arising from contact between sea turtles and marine debris, including petroleum products, are hazards for sea turtles of all ages and usually result in death. Fishing gear has been implicated in many sea turtle deaths. Trawling nets catch turtles and they subsequently drown. Turtles also become entangled in crab pot lines, gill nets, and pound net leader hedging and drown. When they hatch, all sea turtles move instinctively toward the brightest horizon — under natural conditions, the sea. Beach development increases light intensities inland, and many hatchlings die from desiccation, predation, or vehicular traffic as they erroneously travel inland.

Sea turtles have inhabited the earth for 200 million years; they saw the dinosaurs evolve *and* go extinct. It has only been during the last 100 years — and a direct result of human activities — that *all* sea turtle species have been reduced to numbers warranting application of endangered status. It is important to have live and dead sea turtles examined by specialists to aid ongoing assessments of population status and to help identify dangers to sea turtles. If any sea turtle is found, dead or alive, please contact the Virginia Institute of Marine Science Turtle Stranding Network, 804-642-7313, as soon as possible.

Several other factors pose additional threats. Acid rain that reduces the pH of aquatic systems and forest soils has harmed anuran tadpoles and salamander larvae and altered the distributions of terrestrial salamanders. Two introduced species are potentially dangerous to some of Virginia's rare species. Defoliation caused by the gypsy moth could alter the thermal regime of the hardwood forest harboring rare terrestrial salamanders of the genus *Plethodon*. Tiger salamander larvae from midwestern populations sold as waterdogs for fish bait could be released into Virginia's waters, where they could encounter a natural population of the species, which is endangered in Virginia. Because the same species is involved, breeding could easily take place. The danger is that midwestern salamanders may introduce genes into a Virginia population that could cause an adverse affect, such as a change in the response to cues for reproductive migration or an alteration of the tolerance of local soil or water chemistry. The actual effects of introduced species on any rare amphibian or reptile has not yet been evaluated in Virginia.

To maintain the long-term viability of the rare amphibians and reptiles of Virginia it is necessary to curtail the collecting of individuals, to protect populations and their habitats, and to prevent community-wide catastrophes. We cannot protect these species without first protecting the places where they live, their habitats. Conservation biologists are becoming increasingly aware that the protection of large portions of the natural landscape is the only real answer to conserving biodiversity for the long term. A totally integrated conservation approach is necessary in order to understand how the various threats influence each species. Without solid information on the biology of the species at the individual and population levels, and on all pertinent aspects of their environment, we cannot write and implement effective management plans or effectively protect the species from further decline.

Abridged from "Amphibians and Reptiles" by Joseph C. Mitchell, pp. 411–422 in *Virginia's Endangered Species*, coordinated by Karen Terwilliger, 1991 (Blacksburg, VA: The McDonald & Woodward Publishing Company).

Barking Treefrog

Hyla gratiosa

THREATENED

DESCRIPTION: The barking treefrog is the largest species of *Hyla* in the United States, reaching a snout to vent length of about 3 inches. The arms are thick and muscular with a distinctive fold of skin on the wrists. The large hands and feet terminate in large, wide toe pads and are extensively webbed. The skin is somewhat glandular. Coloration is usually bright green; large brown to black spots may or may not be present. A white or yellow lateral stripe occurs but may be broken posteriorly. The mature tadpole is characterized by a very high dorsal fin, dark saddles, and a long, acuminate tip of the tail which terminates in a flagellum. The breeding call is a single loud, clear *tonk,* repeated approximately once per second. A barking call is made from trees near breeding sites. From a distance, either call resembles that of a barking or baying dog. **DISTRIBUTION:** The barking treefrog occurs on the Coastal Plain and adjacent parts of the Piedmont from Mathews County southward. **HABITAT:** In Virginia, choruses of the barking treefrog gather at temporary ponds beneath open canopied forests. **LIFE HISTORY:** The barking treefrog is both arboreal and fossorial. It may dig into the ground during the cold season or in exceptionally dry weather. It feeds on arboreal insects. Breeding occurs in the late spring or summer following heavy rain. Choruses usually do not number more than 20 to 25 males that call from a floating position. **THREATS:** The small number of males calling at any one site makes them vulnerable to overcollecting. The major threat, however, is the continued logging of stands of natural forests. **NOTES:** Barking treefrogs are capable of altering their color. A frog that is bright green and spotless one moment may be heavily spotted or dark green brown half an hour later. Virginia's populations are relicts of a range that was formerly more extensive in the Middle Atlantic region.

Abridged from "Barking Treefrog" by Christopher A. Pague and David A. Young, pp. 426–427 in *Virginia's Endangered Species,* coordinated by Karen Terwilliger, 1991 (Blacksburg, VA: The McDonald & Woodward Publishing Company).

Mabee's Salamander

Ambystoma mabeei

ENDANGERED

DESCRIPTION: Mabee's salamander is a small, stout, mole salamander with a relatively small head and a maximum total length of a little less than 5 inches. The tail comprises about 68 percent of the total length and is compressed in the distal half of its length. Toes are characteristically long. Body color is dark brown to grayish brown above and paler below. Whitish flecks occur on the sides and may be so abundant as to form a mottled pattern. DISTRIBUTION: Mabee's salamander is known from the City of Suffolk and the counties of Isle of Wight, Gloucester, Southampton, and York. HABITAT: Breeding sites in Virginia are fish-free vernal pools on the Coastal Plain. Surrounding forests are generally composed of bottomland hardwoods mixed with pine. LIFE HISTORY: Little is known of the life history of Mabee's salamander. The larval period is spent in ponds or pools. Metamorphosis occurs in April and May when juveniles leave the ponds for a territorial life, returning only to breed. THREATS: Drainage and other serious alterations of vernal ponds destroy breeding habitats or otherwise create unsuitable conditions. Most threats are from urbanization and some forestry practices. NOTES: Mabee's salamander is restricted to lower elevations on the Coastal Plain of Virginia and the Carolinas. The species is not uncommon in the York County sites, but very few specimens are known from the other sites.

Abridged from "Mabee's Salamander" by Christopher A. Pague and Joseph C. Mitchell, pp. 427 and 429 in *Virginia's Endangered Species,* coordinated by Karen Terwilliger, 1991 (The McDonald & Woodward Publishing Company).

Eastern Tiger Salamander

Ambystoma tigrinum tigrinum

ENDANGERED

DESCRIPTION: The tiger salamander is a robust mole salamander with a broad depressed head and widely separated eyes. The tail comprises 45 to 51 percent of total length and is compressed along most of its length. The general color is dark brown to dull black with irregularly spaced and variously shaped yellow to olive blotches on the back. Yellow and olive dominate the chin and throat. The underside is yellow with dark blotches. Maximum total length is about 10 inches. DISTRIBUTION: The eastern tiger salamander is known from Mathews, York, and Augusta counties. HABITAT: The terrestrial habitat of the eastern tiger salamander is essentially any substrate suitable for burrowing. Breeding habitats include sinkhole ponds and vernal pools. Ponds with fish are unsuitable. LIFE HISTORY: Tiger salamanders spend their larval period in freshwater environments and most of their adult life in terrestrial burrows. Both larvae and adults are voracious predators, feeding on insects, crustaceans, amphibian larvae, and mollusks. Adults migrate to breeding sites in late winter. Eggs masses may be attached to submerged twigs or grasses, or to the bottom substrate. THREATS: Primary threats are modifications to the breeding sites such as draining, fish introductions, and succession.

Alteration of local hydrology may decrease the time water is available in ponds and thus may reduce larval survivorship. A threat of genetic contamination from introduced stock used as fish bait (waterdogs) is also a possibility. NOTES: The eastern tiger salamander is common throughout much of its range in the eastern United States but extremely rare in Virginia. Although known from four localities in this state, it is probably still extant in only two.

Abridged from "Eastern Tiger Salamander" by Christopher A. Pague and Kurt A. Buhlmann, pp. 431–433 in *Virginia's Endangered Species*, coordinated by Karen Terwilliger, 1991 (Blacksburg, VA: The McDonald & Woodward Publishing Company).

Shenandoah Salamander

Plethodon shenandoah

ENDANGERED

DESCRIPTION: The Shenandoah salamander is a slender, moderate-sized
salamander with a total length of 3½ to
4½ inches in adults. The body is dark
brown, with two color phases. The striped color phase has a narrow red to yellow stripe
down the center of the back. In the unstriped phase, the back is uniformly dark brown
with scattered brass colored flecks. In both color phases, white or yellow spots occur along
the sides. DISTRIBUTION: The Shenandoah salamander is only known from The Pin-
nacle, Stony Man Mountain, and Hawksbill Mountain in the Blue Ridge Mountains. All
three populations lie within Shenandoah National Park. HABITAT: The Shenandoah
salamander is found only in north and northwest facing talus slopes at elevations greater
than 2,900 feet above sea level. Within the talus, it is found in the soil that has accumulated between and under the rocks. LIFE HISTORY: This species is an opportunistic
predator, consuming any invertebrate small enough to ingest, particularly insect larvae
and mites. As with other species of the genus *Plethodon*, it is entirely terrestrial with no
aquatic larval stage. THREATS: The primary threat is habitat deterioration through natural
succession as surrounding woodland encroaches upon the talus slopes.. The loss of appropriate talus conditions allows the related red-backed salamander, *Plethodon cinereus*, a
woodland species, to enter the talus and displace the Shenandoah salamander. Drought
may also be a factor due to the limited habitat available within the talus. Defoliation of
the overstory by gypsy moths
could increase exposure to
sunlight in the talus and increase evaporation rates to the
detriment of the Shenandoah
salamander. Soil acidification
may also be a threat. NOTES:
Because the primary threat is
a natural one, little can be
done to protect this species
other than reduce human intrusion into its habitat. The
Shenandoah salamander was
federally listed as endangered
in 1989. It had been listed as
endangered by the state of
Virginia in 1987.

Abridged from "Shenandoah Salamander" by Addison H. Wynn, pp. 439–442 in *Virginia's Endangered Species*, coordinated by
Karen Terwilliger, 1991 (Blacksburg, VA: The McDonald & Woodward Publishing Company).

Loggerhead Sea Turtle

Caretta caretta

ENDANGERED

DESCRIPTION: In Virginia's waters, loggerhead carapace lengths range from 8 inches to more than 48 inches and, except for hatchlings which may be found on the Atlantic Coast during the hatching season, weigh 10 to 65 pounds. The back of the carapace and appendages is mahogany to reddish brown, and the plastron and underside of the appendages are cream-yellow. The loggerhead is the only reddish brown sea turtle in our waters. The carapace of the loggerhead is distinctly heart-shaped, while those of the ridley and green sea turtles are round. **DISTRIBUTION:** The most abundant sea turtle in the waters of Virginia, loggerhead turtles are found in Chesapeake Bay, in the estuarine parts of all major tributaries of the Bay, along Virginia's Atlantic coast, and into the channels and lagoons between and landward of the barrier islands. Loggerheads nest in small numbers on the Atlantic Coast of Virginia, most commonly at Virginia Beach. **HABITAT:** Loggerheads reside in deeper channels, usually at river mouths or in the open Bay. They are normally found in Virginia's waters from May through November. **LIFE HISTORY:** Newly hatched loggerheads enter the Gulf Stream and its associated currents, where they find food and refuge within floating mats of *Sargassum*. Juveniles make one or more trips around the North Atlantic gyre, after which they depart from their pelagic existence and enter inshore habitats during the summer months. While in Chesapeake Bay, loggerheads drift passively with the tides within a relatively restricted range, foraging on their preferred prey, horseshoe crabs, *Limulus polyphemus*. The turtles depart the Bay at the onset of cold weather and travel south along the coast past Cape Hatteras. From there their movements have not been tracked; some may winter in the Gulf Stream off North Carolina; others may winter as far south as Florida. **THREATS:** If nesting females are disturbed before egg laying has begun, the nesting attempt is usually abandoned. Vehicular traffic, habitat destruction, and nest predation by raccoons, ghost crabs, and feral animals take a toll on nests. Many hatchlings die from desiccation, predation, or vehicular traffic as they travel inland toward the artificial light of beachside developments. Entanglement, ingestion, and other complications arising from contact between sea turtles and marine debris and fishing nets are hazards for sea turtles of all ages. In Chesapeake Bay, many loggerheads are killed each year from wounds caused by boat propellers. **NOTES:** The loggerhead was federally listed as endangered in 1978.

Abridged from "Loggerhead Sea Turtle" by John A. Keinath and John A. Musick, pp. 445–448 in *Virginia's Endangered Species*, coordinated by Karen Terwilliger, 1991 (Blacksburg, VA: The McDonald & Woodward Publishing Company).

Atlantic Green Sea Turtle

Chelonia mydas mydas

THREATENED

DESCRIPTION: Green sea turtles found in Virginia's waters have carapace lengths ranging from 9 to 20 inches and weigh less than 9 pounds. The back of the carapace and appendages is dark green to brown, often with lines radiating from the posterior margin of each carapacial scute. The plastron and underside of appendages are cream white. DISTRIBUTION: Although historically reported as abundant in Virginia's waters, only two live individuals were reported from 1979 to 1989, one from the York River and one from the Potomac River. Another individual was reported from the Potomac River in 1993. HABITAT: Nonmigrating green turtles prefer sea grass flats such as occur in shallow areas of the Chesapeake Bay. Juvenile green turtles can be found in temperate areas, but adults — noted for their long migrations and remarkable navigational skills — are strictly tropical. LIFE HISTORY: Green sea turtles nest on tropical beaches of the Gulf of Mexico and Caribbean Sea, as well as on the Atlantic Coast of Florida. Hatchlings take refuge in weedlines in the open ocean and travel with the currents. Juveniles are reportedly omnivorous, whereas adults are herbivorous and feed primarily on vascular sea gasses. Stomach contents of individuals stranded in Virginia included both eelgrass, *Zostera*, and the macroalgae sea lettuce, *Ulva*. THREATS: As with other sea turtle species, major adverse impacts on green sea turtle numbers occur on the nesting beaches. At sea, small individuals are consumed by large fish; both juveniles and adults die as a result of various commercial fishing practices. In Virginia, some dead green sea turtles exhibited constriction marks on their appendages which suggest entanglement in fishing gear. NOTES: Increases in the numbers of juvenile green sea turtles collected from Virginia's waters each year may indicate successful conservation efforts in some tropical nesting areas. Improved water quality in Chesapeake Bay and its relationship to the resurgence of sea grass beds in the Bay may also be contributing to the recent increase in green turtle sightings. The green turtle's common name is derived from the color of the fat (calipee) inside the shell, not from external coloration. Calipee is the principal ingredient in clear turtle soup, the market for which has caused the extinction of many green turtle populations. The green sea turtle was federally listed as threatened in 1973.

Abridged from "Atlantic Green Sea Turtle" by John A. Keinath and John A. Musick, pp. 448–450 in *Virginia's Endangered Species*, coordinated by Karen Terwilliger, 1991 (Blacksburg, VA: The McDonald & Woodward Publishing Company).

Atlantic Hawksbill Sea Turtle

Eretmochelys imbricata imbricata

ENDANGERED

DESCRIPTION: The carapace and upper sides of appendages are an attractive combination of amber, brown, and black. The plastron and underside of appendages are yellow, often with dark brown or black spots in young individuals. The only hawksbill recorded in Virginia's waters had a carapace length of 12 inches and weighed less than 2 pounds. DISTRIBUTION: Hawksbill sea turtles are usually restricted to tropical oceans, but individuals have been found from southern Brazil to New England. A live specimen — probably a lost waif — caught incidentally by a waterman clamming at the mouth of the James River in November 1990 confirmed for the first time that this species is a member of Virginia's fauna. HABITAT: Hatchlings probably take up a pelagic existence in *Sargassum* mats in major ocean currents. Hawksbill turtles are primarily coral reef dwellers; any found in Virginia's waters are most likely lost waifs. LIFE HISTORY: Hawksbill sea turtles nest on tropical beaches of the Gulf of Mexico and Caribbean Sea, and hatchlings probably take refuge in weedlines in the open ocean and travel with the currents. They feed primarily on sponges. THREATS: As with other sea turtles, by far the greatest threat to hawksbill turtles is humans. Small sizes are stuffed as curios, the larger individuals are sought for the translucent scutes of the carapace (tortoise shell) to be made into jewelry and other items. As with all species of sea turtle, contact with marine debris is a growing concern. NOTES: The common name of the hawksbill sea turtle is derived from the shape of its beak. Because of the trade in hawksbill items, the species is in danger of becoming extinct. It was federally listed as endangered in 1973.

Abridged from "Atlantic Hawksbill Sea Turtle" by John A. Keinath and John A. Musick, pp. 450–451 in *Virginia's Endangered Species*, coordinated by Karen Terwilliger, 1991 (Blacksburg, VA: The McDonald & Woodward Publishing Company).

Kemp's Ridley Sea Turtle

Lepidochelys kempii

ENDANGERED

DESCRIPTION: Those ridleys found in Virginia's waters are juveniles with a carapace length of 8 to 23 inches and a weight less than 10 pounds. The back of the carapace and appendages is charcoal grey to drab olive green, and the plastron and undersides of appendages are white. DISTRIBUTION: Kemp's ridley is found along the Atlantic Coast of Virginia and throughout the lower Chesapeake Bay. HABITAT: In Chesapeake Bay, Kemp's ridleys are found in shallow, nearshore sea grass beds, especially where their preferred food, blue crabs, *Callinectes sapidus*, are found. They are usually in Virginia's waters from May through November. LIFE HISTORY: The nesting grounds of this turtle were unknown to science until 1963, when a home film taken in 1947 revealed more than 40,000 ridleys of both ridley species congregating on one mile of beach in Tamaulipas, Mexico, the only major nesting area. The *arribada* or mass nesting of the ridleys is unique. (Presently fewer than 500 females nest on that beach.) Hatchlings adopt a pelagic existence in weedlines of the major currents of the Gulf of Mexico and North Atlantic Ocean. Juveniles utilize estuaries such as Chesapeake Bay for foraging during warmer months. Ridleys eat benthic invertebrates, primarily blue crabs, in Virginia's waters. THREATS: Egg collecting, predation, and slaughter for meat and leather were once commonplace, but joint protection by Mexico and the US has greatly reduced those threats. Capture and subsequent drowning in shrimp trawls is now the major threat to the continued existence of this species. As with the other species of sea turtles, contact with marine debris is a growing concern. Ridleys from Virginia's waters have been injured or killed by boat propellers, drowned as a result of entanglement in debris or stationary fishing gear, or drowned in the autumn bottom fishery off the coast. NOTES: An unusual way to distinguish live ridleys from other live sea turtles has been described. A rap with the knuckles on the carapace produces a hollow sound, like thumping a hollow log; other species produce a solid sound. Kemp's ridley was federally listed as endangered in 1973.

Abridged from "Kemp's Ridley Sea Turtle" by John A. Keinath and John A. Musick, pp. 451–453 in *Virginia's Endangered Species*, coordinated by Karen Terwilliger, 1991 (Blacksburg, VA: The McDonald & Woodward Publishing Company).

Leatherback Sea Turtle

Dermochelys coriacea

ENDANGERED

DESCRIPTION: The leatherback is the world's largest sea turtle and is rarely confused with other sea creatures. Specimens from Virginia havae been estimated to weigh 400 to 800 pounds; carapace lengths range from 4 to 6 feet. The carapace and body have no horny scutes but are covered with a smooth, delicate skin that looks and feels rubbery or leathery. The carapace has seven longitudinal ridges, and the body is black on the upperside with white, yellow, or pink undersides. Deteriorating specimens can be distinguished from other sea turtles by their large size and lack of bony carapace or claws on the flippers. **DISTRIBUTION:** Leatherbacks are routinely seen off the mouth of Chesapeake Bay, where they presumably feed on jellyfish washing out of the Bay. Leatherbacks have been observed in Chesapeake Bay as far north as the Patuxent River. **HABITAT:** Leatherbacks, the most pelagic of the sea turtles, forage in coastal and offshore waters but occasionally wander close to shore and into estuaries. Leatherbacks occur in Virginia's waters primarily during warm months (May through September). **LIFE HISTORY:** The leatherback is the only extant turtle known to be warm-blooded, a trait which permits its survival in cool waters as far north as Canada. The western Atlantic population nests primarily on Caribbean shores. Small juveniles are rarely observed, thus little is known of their habits. Leatherbacks feed on soft-bodied pelagic invertebrates such as jellyfish. In Virginia, their diet most likely consists of sea nettle, *Chrysaora quinquecirrha*, and moon jellyfish, *Aurelia aurita*. **THREATS:** As with other sea turtles, major impacts on leatherbacks are the collection of newly deposited eggs, nest predation, and the killing of nesting females. In the waters of the US, deaths are attributed to intentional mutilation by gunshots, apparently for sport; entanglement in fishing gear or debris; and curiosity killing. Plastics, which may resemble jellyfish, are eaten by leatherbacks and have been implicated as a cause of death by impaction of the gut. **NOTES:** Although leatherbacks were reportedly once common visitors to Chesapeake Bay, very few are now sighted within the Bay. The Chesapeake Bay Bridge-Tunnel may be a physical deterrent. The leatherback sea turtle was federally listed as endangered in 1973.

Abridged from "Leatherback Sea Turtle" by John A. Keinath and John A. Musick, pp. 453–455 in *Virginia's Endangered Species*, coordinated by Karen Terwilliger, 1991 (Blacksburg, VA: The McDonald & Woodward Publishing Company).

Wood Turtle

Clemmys insculpta

THREATENED

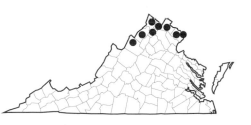

DESCRIPTION: Adult wood turtles have carapace lengths of 5 to 10 inches. The broad, low-keeled carapace is roughened with pyramids of concentric growth annuli. The carapace is gray to brown, often with black and yellow lines radiating from the upper posterior corners of each pleural scute. A dark oblong blotch occurs in the center of each scute of the hingeless yellow plastron. The large head is black with a non-projecting snout and a notched upper jaw. Yellow, orange, or red pigment usually occurs on the neck and upper forelegs. Hatchlings are grayish brown and lack any orange or red on the neck or legs. Wood turtles lack the black, yellow-spotted carapace of the spotted turtle, *Clemmys guttata*, with which they occur in northern Virginia. **DISTRIBUTION:** In Virginia, the range extends from Arlington and northern Fairfax counties westward through Loudoun and Clarke counties to Frederick, Warren, Shenandoah, and Rockingham counties. **HABITAT:** The wood turtle is found primarily in and near clear brooks and streams in deciduous woodlands. It hibernates in deep pools or under the mud or sand bottom of its waterway, just sitting on the bottom, or under the overhanging roots of trees along the bank. Some have been found under submerged logs, in beaver lodges, and in muskrat burrows. **LIFE HISTORY:** Wood turtles are active from April through November. Both sexes wander extensively on land until autumn, when they return to aquatic hibernation sites. Although highly terrestrial, wood turtles must remain in moist habitats. The wood turtle is omnivorous. Animal foods include a wide variety of insects and mollusks, earthworms, dead fish, tadpoles, newborn mice, and other turtles' eggs. Mating occurs between April and September, usually in water. Nesting normally occurs in June. **THREATS:** Rapid residential and commercial development in northern Virginia is destroying much of the habitat of the wood turtle. The species is also very popular in the pet trade, and some colonies could be decimated as a result of collecting for this purpose. **NOTES:** The wood turtle ranges from Cape Breton Island, Nova Scotia, to northern Virginia. The species is protected as an endangered species in Michigan, New York, and Wisconsin.

Abridged from "Wood Turtle" by Carl H. Ernst and John F. McBreen, pp. 455–457 in *Virginia's Endangered Species*, coordinated by Karen Terwilliger, 1991 (Blacksburg, VA: The McDonald & Woodward Publishing Company).

Bog Turtle

Clemmys muhlenbergii

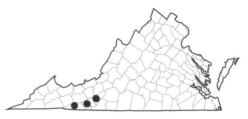

ENDANGERED

DESCRIPTION: The bog turtle is a small freshwater turtle reaching a maximum carapace length of 4½ inches. The surface of the carapace is roughened with growth annuli, often worn smooth on the shells of old adults, and black to brown or mahogany in color. A low medial keel is present in juveniles, but is often worn smooth with age. The hingeless plastron is usually black with irregularly shaped blotches of yellow to cream along the midline. The skin on the head, neck, and limbs is brown to pinkish brown; there may be some red mottling on the limbs. A large conspicuous orange, yellow, or red blotch lies behind each eye, but this may degenerate in old adults. Shell and skin patterns in juveniles differ slightly from adults. In juveniles, the carapace is round and dark brown in color, and the plastron is yellow with a large black blotch in the center. The bright orange patch behind the eye is present at hatching. DISTRIBUTION: Bog turtles are found in Virginia only on the southern Blue Ridge Plateau. HABITAT: In Virginia, bog turtles are found in upland freshwater wetlands characterized by open fields, meadows, marshes with slow moving streams, ditches, and boggy areas. In July and August these turtles aestivate in soft mud. During the winter they hibernate below the frost zone in holes, muskrat burrows, clumps of sedges, or the mud of waterways. LIFE HISTORY: Nothing is known of the diet of Virginia's bog turtles; elsewhere they are omnivores. Mating occurs from late April to early June. Eggs are laid from May to early July in shallow nests in grass tussocks, moss, or soft soils. Most hatching occurs in August, but some young do not emerge until early October or the following May or April. THREATS: Habitat loss from the drainage of wetlands for development and agriculture poses the primary threat to bog turtles. Also, the species commands high prices in the pet trade, and many turtles have been removed from populations throughout the range for this purpose. NOTES: The bog turtle is one member of a unique biological assemblage on the southern Blue Ridge Plateau. It was listed as endangered in Virginia in 1987.

Abridged from "Bog Turtle" by Joseph C. Mitchell, Kurt A. Buhlmann, and Carl H. Ernst, pp. 457–459 in *Virginia's Endangered Species*, coordinated by Karen Terwilliger, 1991 (Blacksburg, VA: The McDonald & Woodward Publishing Company).

Eastern Chicken Turtle

Deirochelys reticularia reticularia

ENDANGERED

DESCRIPTION: The eastern chicken turtle is a moderate-sized freshwater turtle reaching a maximum carapace length of 10 inches. The carapace is elongated, somewhat high-domed, and its surface is roughened with numerous longitudinal ridges or striations. The head and neck are long, nearly the length of the carapace when fully extended. The carapace is brown to olive with a yellow to light brown netlike pattern. Black spots may be present on the ventral side of the marginal scutes. The plastron is usually plain yellow. Thin yellow stripes occur on the black skin of the neck and there are yellow and black stripes on the rear of the thighs. DISTRIBUTION: The only known population of chicken turtles in Virginia is at Seashore State Park in the City of Virginia Beach. HABITAT: The chicken turtle occupies the freshwater ponds located along the forested dunes in Seashore State Park. They do not abandon aquatic habitats even when the ponds are nearly dry in drought years. LIFE HISTORY: Chicken turtles are basking turtles and are sometimes seen on logs and stumps. They are omnivorous and eat crayfish, tadpoles, and aquatic plants. THREATS: An important threat to this species is the very small population size; fewer than 10 individuals exist in the Park. Recruitment levels are dangerously low, probably due to both recurring drought and predation by raccoons and snapping turtles. NOTES: The chicken turtle ranges from the southern edge of Chesapeake Bay south to Florida and thence westward to Texas, Arkansas, and Oklahoma. The nearest population to that of Seashore State Park is at Nag's Head Woods, North Carolina. This species was officially listed as endangered in Virginia in 1987.

Abridged from "Eastern Chicken Turtle" by Joseph C. Mitchell and Kurt A. Buhlmann, pp. 459–461 in *Virginia's Endangered Species*, coordinated by Karen Terwilliger, 1991 (Blacksburg, VA: The McDonald & Woodward Publishing Company).

155

Canebrake Rattlesnake

Crotalus horridus atricaudatus

ENDANGERED

DESCRIPTION: The canebrake rattle-snake is a large venomous snake reaching a maximum length of over 6 feet. The body color is pinkish, gray, yellow, or light brown with a series of dark brown to black chevrons. A brown or chestnut mid-dorsal stripe is present on most individuals, as is a yellowish brown stripe that runs from the eye to the back of the jaw on each side. The underside is cream in color and may be lightly peppered with black; tail is black. One or more loose, keratinazed segments on the base of the tail form a rattle. A deep pit is located between the eye and nostril on each side. The mountain counterpart of the canebrake, the timber rattlesnake, *Crotalus horridus horridus*, lacks the short eye-to-jaw stripe, the chestnut mid-dorsal body stripe, and the pink coloration. **DISTRIBUTION:** Most of the known localities in Virginia are in the lower Peninsula and in the southeastern corner of the state east of the Dismal Swamp. **HABITAT:** Canebrake rattlesnakes occupy hardwood and mixed hardwood-pine forests, canefields, and the ridges and glades of swampy areas. Preferred habitat is mature hardwood forests containing numerous logs and a layer of leaves and humus. Canebrakes overwinter singly or in small numbers in the base of hollow trees or in stumps. **LIFE HISTORY:** Canebrakes are primarily predators of small mammals but will consume other vertebrates. They bear live young in litters of 7 to 13. Maturity is reached at six years of age and at about 40 inches snout to vent length in females, and four years of age and 36 inches snout to vent length in males. **THREATS:** The primary threat to canebrake populations in Virginia is loss of habitat from urban development. **NOTES:** High rates of habitat loss mean that those populations still remaining north of the James River and most of those in southeastern Virginia will probably be extirpated by the year 2000. There are only four areas where viable populations are likely to remain: Dismal Swamp National Wildlife Refuge, Northwest River Park, riverine habitat between these locations, and an area north of Dismal Swamp.

Abridged from "Canebrake Rattlesnake" by Joseph C. Mitchell and Don Schwab, pp. 462–464 in *Virginia's Endangered Species*, coordinated by Karen Terwilliger, 1991 (Blacksburg, VA: The McDonald & Woodward Publishing Company).

Eastern Glass Lizard

Ophisaurus ventralis

THREATENED

DESCRIPTION: The eastern glass lizard reaches a maximum total length of about 43 inches. The slender, serpentine body lacks legs. The tail, when complete, is more than twice as long as the body. The dorsal ground color is olive brown. A broad tan dorsal stripe extends onto the tail. A dark stripe of greenish black peppered with white occurs on each side above the lateral groove. The underside of body and tail is white. Three to seven short white bars bordered with black occur on the side of the neck and head. DISTRIBUTION: In Virginia this species has been found only on the Currituck Spit in Back Bay National Wildlife Refuge and in False Cape State Park, City of Virginia Beach. HABITAT: Eastern glass lizards inhabit pine flatwoods, mesic hammocks, wet meadows, and damp grassy areas. LIFE HISTORY: Although little is known of their life history, glass lizards are probably diurnal. They spend most of their lives underground and under vegetation. Grasshoppers are primary prey, but their diet also consists of spiders, crickets, cockroaches, beetles, caterpillars, snails, other lizards and small snakes. Females coil around their eggs in nests which are depressions under objects on the ground. Females do not defend their eggs as do skinks, but will gather them if they become scattered. Brooding females leave if the nest cover is removed. THREATS: The primary threat in Virginia is loss of habitat due to planned construction activities at both False Cape State Park and Back Bay National Wildlife Refuge. Lack of maintenance of natural grassy areas may also be detrimental. NOTES: Eastern glass lizards are distributed across much of the Coastal Plain of southeastern North America. The first record of the species in Virginia was made in 1981.

Abridged from "Eastern Glass Lizard" by Joseph C. Mitchell and Christopher A. Pague, pp. 464–466 in *Virginia's Endangered Species*, coordinated by Karen Terwilliger, 1991 (Blacksburg, VA: The McDonald & Woodward Publishing Company).

Birds

Birds represent a highly mobile component of Virginia's fauna. The avifauna of the state consists of 422 species: 359 occur regularly, 41 are accidental, 20 are hypothetical, and two are extinct. Many species occur in all of the five physiographic provinces; others are restricted to only one or some of the provinces. Of the 359 regularly occurring here, 214 have bred in the state. Many of these breeding species have a wide distribution on the North American continent. Several, reaching the limits of their breeding ranges here, are uncommon in Virginia. Some of the species that occur in Virginia but do not breed here are winter residents or transients. No bird species is endemic to Virginia.

For some birds, detailed distributional and breeding data are poorly known. The Virginia Breeding Bird Atlas Project, co-sponsored by the Virginia Department of Game and Inland Fisheries and the Virginia Society of Ornithology, has recently filled many gaps in our knowledge of breeding bird distributions. In addition, researchers from the Virginia Society of Ornithology, US Fish and Wildlife Service, Virginia Natural Heritage Program, and universities across the state continue to contribute important data on the state's varied avifauna.

Concerns about changes in the populations of North American birds have been voiced for at least two decades. Several independent, broad-scale surveys — the roadside Breeding Bird Survey, protracted breeding bird censuses, and annual banding efforts — have recently indicated widespread, long-term population declines for many migratory bird species. Ornithologists are especially concerned about neotropical migrant land birds, those species that breed in North America and then overwinter in Central and South America and the islands of the Caribbean. The Breeding Bird Survey showed that of the 62 neotropical migrant species, 48 had stable or increasing populations across North America from 1965 to 1979, but from 1978 to 1987, 44 of the same 62 species showed declines in population levels. For some species there are local, regional, and temporal differences in population changes.

In Virginia, at least 84 observers have counted breeding birds along some 46 routes of the Breeding Bird Survey. The surveys have shown changes in the populations of some species of small land birds from 1966 to 1989 — the bank swallow (decreasing), cliff swallow (increasing), Bewick's wren (decreasing), and loggerhead shrike (decreasing). In the 1980s, the bank swallow, Bewick's wren, and Bachman's sparrow were no longer reported on the survey routes in Virginia. For 50 species of land birds that winter in the neotropics, Breeding Bird Survey data from 1980–1989 showed that 14 were declining, 10 were increasing, and 26 showed no significant changes in population levels.

On the other hand, other techniques have revealed nearly stable populations at other sites. Breeding bird census data from Mountain Lake Biological Station in southwest Virginia show that populations of most species — both migrants and nonmigrants — changed very little over the last 20 years. Recruitment of breeding individuals from the extensive, contiguous forest probably offsets losses on the census site from heavy predation by various mammals and nest parasitism by the brown-headed cowbird.

Some researchers believe that the major cause of declining bird populations is forest fragmentation on North American breeding grounds, including the secondary effects of the reduced area of extensive forested tracts, increased forest edge, and subsequent increases in nest predation and cowbird parasitism. Other scientists believe the effects of massive deforestation and forest conversions on the tropical wintering grounds are the major factors. The degree and impact of modification of the habitats of a variety of long-distance migrants on the breeding grounds, the wintering grounds, and along migration routes are subjects of continuing research. Specific causes of population changes in Virginia, whether for short or long distance migrants, are poorly documented, but it appears that habitat destruction and alteration have played, and continue to play, major roles. The need remains to document quantitatively the effects of weather, human disturbance, predation, cowbird parasitism, and environmental contamination.

There probably are many reasons why the populations of any given bird species have declined within the state and across the continent to the point that the bird is now assigned endangered or threatened status. Most reasons relate directly to the manipulation of natural ecosystems by humans. The factors which relate to changing dynamics of populations may be categorized into two groups:

• those that reduce survivorship in the population (for example, overharvesting, increased rates of predation and nest parasitism, reduction of suitable habitat); and

• those that reduce fecundity in the population (for example, competition for nesting sites and reproductive dysfunctions caused by environmental contaminants).

Among the various environmental alterations responsible for the decline of a number of species in Virginia, the modification and destruction of habitats and the presence of environmental contaminants, especially synthetic pesticides, are particularly significant. Recovery of any endangered or threatened species will be contingent upon the degree to which these detrimental factors can be reduced or eliminated.

In Virginia and neighboring states, human population growth is particularly high in the Chesapeake Bay basin. And undeveloped land is being converted for development at least twice as fast as the human population is increasing. Many of the birds of the Coastal Plain of Virginia depend on wetlands for their survival. Between the mid-1950s and the late 1970s, more than 2,800 acres of wetlands were lost annually in Virginia, principally to urban development and agriculture. Nontidal wetlands are disappearing at an even more rapid rate than tidal wetlands.

As the human population of the Chesapeake Bay basin grows, there will be concomitant demands for additional recreational areas and resources, creating further pressures on such sensitive but relatively protected areas as the barrier islands. Much of the development in the Chesapeake Bay basin is in optimal bald eagle habitat; bald eagle nesting, roosting, and feeding habitat continues to give way to housing, highways, parks, airports, and public utilities. Approximately 10 to 15 percent of bald eagle nests in Virginia are adversely impacted each year by development.

Development activities also fragment habitats. Large blocks of once contiguous habitat, such as hardwood forests, are broken up by conversion to agricultural fields and housing developments. Many species of birds occupy very specialized habitats that often, because of these changes, are now found only in small tracts widely separated from each other. Some birds are sensitive to specific minimum area thresholds. A variety of factors, including reproductive characteristics and competition for food and nest sites may be at work in determining minimum area requirements. Decreasing numbers of some species of birds, particularly passerines, can be correlated with decreasing sizes of blocks of suitable habitat. It seems clear that unless present land use trends in Virginia are reversed, populations of many species in the state can be expected to decline as a result of continued habitat fragmentation.

A third critical problem involving habitat is the timber management system which results in tree monocultures. Pine monocultures now cover hundreds of thousands of acres in Virginia on corporate, government, and private lands. An efficient method for growing and harvesting a large volume of timber, tree monoculture creates an artificial, simplified ecosystem that provides little suitable habitat for most animals including birds.

The adverse effects on birds of environmental contaminants, such as synthetic pesticides, are well documented in the scientific literature. In particular, DDT and its metabolites are known to cause eggshell thinning in raptors such as peregrine falcons and bald eagles, thus lowering reproductive potentials. The peregrine falcon was extirpated as a breeding bird in Virginia in the 1960s. The number of territories occupied by bald eagles in the Chesapeake Bay basin declined from 150–200 in 1936 to 55 in 1970. Productivity in 1962 was 0.2 young per nest, about one-eighth the level of 1936. Concentrations of residues of organochlorines in eagle carcasses in the mid-Atlantic region were among the highest for any region in the United States. Following the ban on the use of DDT in the United States in 1973, the numbers and nesting success of bald eagles in the Chesapeake Bay area have steadily increased. Nests of the bald eagle now number over 150; and, in 1994, the species was downlisted from endangered to threatened.

DDT residues in bird species of eastern North America have declined significantly since about 1970. It appears at this time that DDT and related persistent pesticides do not pose threats to Chesapeake Bay bald eagles and other raptors; however, threats to wildlife species from readily degradable insecticides such as carbamates have not been thoroughly documented or evaluated. These latter chemicals also are highly toxic and are known to have caused the deaths of 15 bald eagles in the Chesapeake Bay in the last few years.

Most beach-nesting species exist in greatest numbers on the outer barrier islands and on dredge-spoil areas. Because the barrier islands are protected to a

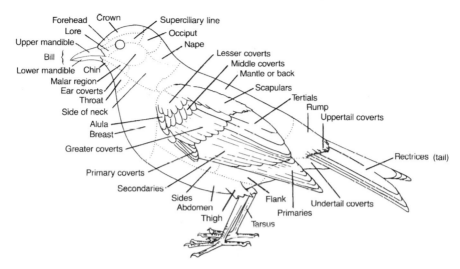

Figure 8. A generalized bird showing selected morphological characters.

great extent by The Nature Conservancy and state or federal governments, they experience little human intrusion and development. These islands, however, are very tenuous and fragile environments and are highly susceptible to erosion and inundation by sea water. In artificially disturbed environments, such as dredge-spoil areas, natural succession of vegetation frequently makes the area unsuitable over time for beach-nesting birds but suitable nesting habitat for species of herons that nest in bushes. Disturbed areas need to be managed properly to maintain them in the most suitable state for species of special concern.

Biologists have sought new and innovative techniques for the management of birds at critically low population levels. The reintroduction of captive-reared peregrine falcons by hacking is an example of an effective management strategy. Fostering and cross fostering of eggs and young are other examples of effective management techniques. Although such practices may be effective in reestablishing or stabilizing a population, they may not bring about a long-term secure future for a species unless the environmental problems which placed it in danger in the first place have been corrected. The pressures on habitats throughout the Commonwealth will be enormous in the future. Our ability and tenacity in planning and utilizing our land resources will surely dictate the future of our avifauna.

Abridged and adapted from "Birds" by Mitchell A. Byrd and David W. Johnston, pp. 477–486 in *Virginia's Endangered Species*, coordinated by Karen Terwilliger, 1991 (Blacksburg, VA: The McDonald & Woodward Publishing Company).

Bald Eagle

Haliaeetus leucocephalus

THREATENED

DESCRIPTION: Adults have white heads and tails, a brownish black body, and yellow bills, eyes, and feet. Immatures are variable in plumage but generally have a dark brown body, tail, and head, each irregularly blotched with cream or white; the bill is brownish, eyes pale yellow gray, and feet lemon yellow. The tail and head become white at age four or five years. Mature birds measure 34 to 43 inches in total length. **DISTRIBUTION:** The bald eagle is usually found near water, but in migration it may occur in almost any part of the state. In Virginia, bald eagles breed primarily in the Chesapeake Bay region, where the population seems to be fairly stable. **HABITAT:** The bald eagle forages along coasts, rivers, and large lakes. Nest sites are found in the midst of large wooded areas adjacent to marshes, on farmland, or in logged-over areas where scattered seed trees remain. Most are remote from human activity and less than a mile from

feeding areas. **LIFE HISTORY:** Virginia's breeding bald eagles appear to be permanent residents. In December or January each pair builds a nest in the fork of a tall tree. Two bluish white eggs are laid between mid-January and mid-March. Most eggs hatch in April and young fledge by early July. Bald eagles are opportunistic feeders and consume a variety of living prey species as well as carrion. They feed primarily on fish either self-caught, scavenged, or pirated from ospreys. Waterfowl (especially those crippled during the hunting season), muskrats, rabbits, squirrels, and road kills are not uncommon in the diet. **THREATS:** The greatest threat in Virginia is urban and residential developments in roosting, foraging, and nesting habitats. **NOTES:** Most of the persistent pesticides which initially caused population declines are no longer in use in North America, and many eagle populations have recovered. As of 1994, the bald eagle is listed both federally and in Virginia as threatened. Previously it had been listed as endangered.

Abridged from "Bald Eagle" by Mitchell A. Byrd, pp. 497–499 in *Virginia's Endangered Species*, coordinated by Karen Terwilliger, 1991 (Blacksburg, VA: The McDonald & Woodward Publishing Company).

Peregrinc Falcon

Falco peregrinus

ENDANGERED

DESCRIPTION: Adult peregrine falcons have long pointed wings, a dark blue or slate back, black top of head and cheeks, which contrast with its white throat and sides of neck. Underparts are white barred with blackish brown. The tail is long, narrow, blue-gray, and rounded, with narrow black bands and a broad subterminal bar tipped with white. In flight, the quick wing beats resemble those of the rock dove (pigeon). Males measure 15 inches; females are larger. DISTRIBUTION: Migrating peregrines pass through Virginia in the first two weeks of October, and a few birds winter on the coast. Recent breeding records in Virginia are all from the Coastal Plain. As reintroduction efforts continue in the mountains, there is a greater chance that resident birds will be found there. HABITAT: The peregrine falcon forages in a wide variety of habitats from coastal waters to open valleys to congested cities. Historically it nested on cliff sites in the Appalachian Mountains. All pairs known to be nesting now (introduced birds) use artificial landscape features such as bridges, towers, multi-story buildings, and even a ship. Transient birds and wintering birds utilize coastal marshes and barrier islands. LIFE HISTORY: Peregrines make no nest of their own but use scrapes in accumulated debris or the abandoned nests of eagles, hawks, and ravens. Clutch size is three or four eggs. Peregrines capture prey by hitting it in the air at the climax of high speed dives clocked at over 100 miles an hour. They eat any small to medium-sized bird, but particularly rock doves, pelagic birds, songbirds, and waterfowl. THREATS: The greatest threat to the reintroduced population is the potential increase of human activity on the barrier islands, where most breeding now occurs. Loss of habitat and human activity in natural habitats in western Virginia have rendered many historic nest sites in the mountains unsuitable. NOTES: The peregrine falcon was a cosmopolitan species. Populations were reduced worldwide during the 1950s and 1960s as a consequence of the use of DDT. The breeding population in North America became restricted to Alaska, Canada, Baja California, and isolated parts of the western US.

Abridged from "Peregrine Falcon" by Mitchell A. Byrd, pp. 499–501 in *Virginia's Endangered Species*, coordinated by Karen Terwilliger, 1991 (Blacksburg, VA: The McDonald & Woodward Publishing Company).

Piping Plover
Charadrius melodus

THREATENED

DESCRIPTION: The piping plover is ghostly in appearance and behavior as it runs over sand, its coloration blending with the background. In flight, a bold white wing bar and largely white tail with dark center can be seen. The pale back, single (often incomplete) black breast band, and yellow-orange legs and feet are distinctive. The stubby bill is orange tipped with black. Average size is about 5½ inches. Piping plovers are often heard before they are sighted. Their ethereal, melodious *peep-peep* or *peep-lo* can be ventriloquial and difficult to pinpoint. On the other hand, the birds may be conspicuous as they feign a broken wing to distract intruders from eggs or young. **DISTRIBUTION:** The piping plover breeds on the barrier islands of the Eastern Shore and near Hampton Roads. **HABITAT:** Piping plovers nest almost exclusively on beaches. A variety of foraging sites are used, including intertidal surf zones, mud flats, tidal pool edges, barrier flats, and sand flats. **LIFE HISTORY:** The nesting season lasts from late April to late July, and only one brood is raised each year. The nest scrape, constructed by the male, is a shallow depression in the sand, typically lined with bits of broken seashells or fine pebbles. Clutch size is four eggs. Precocial chicks all hatch on the same day and leave the nest within hours of hatching. The young immediately forage for themselves, but they are defended by the adults. Piping plovers eat a variety of small invertebrates including marine worms, crustaceans, mollusks, insects, and the larvae and eggs of many small marine animals. **THREATS:** Both avian and mammalian predators are constant threats to eggs and young. These include red foxes, feral cats, herring gulls, fish crows, and grackles. Ghost crabs also prey on eggs and chicks. Increasing development and recreational activities on the barrier islands imperil breeding sites. **NOTES:** The piping plover is endemic to North America, where it breeds in three disjunct populations. In 1985 the US Fish and Wildlife Service listed the piping plover as endangered in the Great Lakes area and threatened elsewhere. Continued protection of the barrier islands is essential to the recovery of this species in Virginia.

Abridged from "Piping Plover" by Robert R. Cross, pp. 501–502 in *Virginia's Endangered Species*, coordinated by Karen Terwilliger, 1991 (The McDonald & Woodward Publishing Company).

Wilson's Plover

Charadrius wilsonia

ENDANGERED

DESCRIPTION: Wilson's plover is distinguished from other plovers in Virginia by its thick black bill. It has a single breast band, white stripe from bill to eye, and anterior portion of crown dark brown to black in breeding males, grayish brown to reddish brown in females, immatures, and winter males. The rest of the breast is white. Legs are dull pink. Total length is about 6¼ inches. **DISTRIBUTION:** Wilson's plover breeds only on barrier islands of the Eastern Shore. It is also a rare transient and summer visitor along the lower Chesapeake Bay and the Atlantic Coast south of Cape Henry. **HABITAT:** Wilson's plovers nest on the upper portions of sandy beaches on barrier islands, usually within 100 feet of dune vegetation, but not in dense vegetation. Often they nest near least terns or piping plovers. Nesting habitat requires both suitable nest areas and nearby suitable foraging sites for chicks, usually mud or sand flats. **LIFE HISTORY:** Wilson's plovers occur in Virginia between 25 April and 5 September. Three pale tan or buff eggs with irregular black speckling are laid in a partially lined scrape in the ground. The parent leads the young to a feeding area within hours of the last egg's hatching, and both parents attend to the young. Adults eat mainly crustaceans, among which fiddler crabs are especially important. **THREATS:** Mammalian and avian predators, human disturbance to nesting birds, and loss of nesting habitat are major threats to Wilson's plover in Virginia. Herring gulls, great black-backed gulls, and fish crows prey on them. Ghost crabs may take some eggs and chicks. **NOTES:** Historical records of distribution and abundance are sketchy but indicate that the numbers of Wilson's plovers in Virginia have declined in this century.

Abridged from "Wilson's Plover" by Peter W. Bergstrom, pp. 502–504 in *Virginia's Endangered Species*, coordinated by Karen Terwilliger, 1991 (Blacksburg, VA: The McDonald & Woodward Publishing Company).

Upland Sandpiper
Bartramia longicauda

THREATENED

DESCRIPTION: The upland sandpiper is a bird of upland grasslands and pastures. It is distinguished by having a thin neck, small head, short bill, relatively dark tail, and yellow legs. Upper parts are mottled or darkly streaked with brown, the throat and breast are noticeably patterned with brown chevrons. After alighting, these sandpipers often briefly hold their wings vertically. The call note is a *t'wit-t'wit*. Total length is 10 inches. **DISTRIBUTION:** The upland sandpiper is a rare or uncommon transient and rare breeder in Virginia, occurring chiefly in five northern counties. The annual state breeding population is about 15 to 20 pairs. Postbreeding and migration records are concentrated at or near breeding locations, on the Eastern Shore, and at a few inland airports. **HABITAT:** Upland sandpipers in Virginia apparently always nest in open farming areas with mixed habitats that include medium-height to tall grasses and fallow or early-stage oldfields. Plowed fields and/or short grass, heavily grazed pastures or sod farms, as well as airports, are also used, especially during migration. **LIFE HISTORY:** Migrant upland sandpipers arrive in Virginia in late March to early April. Nesting begins shortly after birds arrive. The nest is a cup pressed into the roots of a clump of grass or a slight hollow scooped in the ground and lined with fine grasses in dense growth. The four eggs are creamy buff, finely speckled and blotched with light reddish brown or chocolate. Family groups appear to remain together at least until postbreeding migration. Fall migration begins in July and occurs into September. The upland sandpiper's diet consists of grasshoppers, crickets, weevils, flies, worms, spiders, and various grass seeds. **THREATS:** Population levels are dropping throughout the east as preferred field habitats are changed to other land uses, including development. **NOTES:** Market hunters took large numbers of this species in the 1800s, but the species' range apparently expanded westward after hunting was declared illegal. Today the upland sandpiper breeds from central Alaska to northwestern British Columbia, eastward across Canada and the northern US to New England and Virginia. It winters in South America from Surinam and Brazil to Argentina and Uruguay.

Abridged from "Upland Sandpiper" by John B. Bazuin, Jr., pp. 504–505 in *Virginia's Endangered Species* coordinated by Karen Terwilliger, 1991 (Blacksburg, VA: The McDonald & Woodward Publishing Company).

Gull-Billed Tern

Sterna nilotica

Threatened

DESCRIPTION: The gull-billed tern is the whitest of North American terns. The breeding adult has pale gray upperparts, a black crown and nape, and a short, thick, black, gull-like bill. It has a stocky body and moderately forked white tail. These terns have a distinctive call, a dry, raspy *kay-tih-DID* or *kay-DID*. Total length is about 13 inches. **DISTRIBUTION:** In Virginia, gull-billed terns currently breed only on the barrier islands of the Eastern Shore. **HABITAT:** The gull-billed tern was originally a marsh-nester like the Forster's tern and laughing gull. On the Eastern Shore barrier islands, it breeds on higher areas of beaches and over-washes near sand dunes **LIFE HISTORY.** Gull-billed terns return to Virginia nesting sites from late April to early May. Nesting occurs from late May to mid-July. The nest is usually a shallow depression scraped in the sand and lined with bits of shells or dried vegetation. Two to five pink-buff eggs, lightly spotted with dark brown, are laid from June to July. Gull-billed terns are colonial nesters often found in association with other tern species. Unlike most other terns, gull-billed terns usually feed by hawking insects over marshes and croplands. **THREATS:** Current threats to the gull-billed tern population include loss of suitable habitat to beach erosion, development, and urbanization along Virginia's Eastern Shore. Nesting disturbances are also a factor and are likely to increase with increased human recreational activities. **NOTES:** Historically, the gull-billed tern population was decimated by egg-collectors and plume hunters.

Abridged from "Gull-Billed Tern" by Bill Williams, pp. 509–510 in *Virginia's Endangered Species* coordinated by Karen Terwilliger, 1991 (Blacksburg, VA: The McDonald & Woodward Publishing Company).

Roseate Tern

Sterna dougalli

ENDANGERED

DESCRIPTION: This tern is very similar in size and color to the other *Sterna* terns found on the coast of Virginia. Adults have a white ventral surface with an indistinct pinkish coloring on the chest and upper belly. This pink cast can be difficult to see. Bill color (black with little to no red coloring at the base) and the forked, white tail help distinguish the roseate tern. When standing, the tail extends well past the folded wings. This tern is lighter in color than Forster's and common terns; and its legs and feet are black instead of red to orange as in Forster's and common terns. The juvenile's cap extends well onto the forehead. Total length is 15 inches. **DISTRIBUTION:** Formerly breeding on Eastern Shore barrier islands, the roseate tern now is seen only in passage through Virginia and is usually restricted to offshore waters. All recent sightings have been along the Atlantic Coast. **HABITAT:** The roseate tern is restricted to Atlantic coastal environments. Nesting occurs on upper, vegetated portions of sandy beaches. Nests are usually located within thick vegetation, usually under cover. **LIFE HISTORY:** Adults arrive at nesting colonies on the east coast of the US in April. Eggs are laid from April through June in southern areas, with some eggs still present into August in more northerly locations. Incubation averages 21 days. Approximately 28 days after hatching, juveniles are able to fly. The diet is principally small fish such as flounder, herring, sea launce and mullet. The roseate tern winters in the West Indies and along the northern coast of South America. **THREATS:** Current conditions which may prevent the return of the ro-

seate tern as a breeding species in Virginia include loss of habitat due to human development, urbanization, beach stabilization, and erosion. Disturbance of tern nesting colonies and increases in herring and black-backed gull populations may also limit the ability of this tern to become reestablished as a breeder. **NOTES:** Unpublished data show May and August as the months when most sightings of roseate terns occur.

Prepared by Don Schwab, Virginia Department of Game and Inland Fisheries, December 1993.

Red-Cockaded Woodpecker

Picoides borealis

ENDANGERED

DESCRIPTION: The red-cockaded woodpecker has a black and white cross-barred back, black cap and nape, and distinctive white cheeks. The small red cockades on each side of the black cap, present only on males, are difficult to see in the field. Its distinctive, raspy call note can be used to locate birds in the forest.Total length is about 7¼ inches. **DISTRIBUTION:** This species is very rare in Virginia. Only five colonies, all in Sussex County, are currently active. **HABITAT:** Most colonies are found in open, park-like stands of mature long-leaf, loblolly, or shortleaf pine. The red-cockaded woodpecker selects large, 65 to 110 year old living pine trees for excavation. **LIFE HISTORY:** Individual clans or family groups of two to six birds maintain year-round territories near their nesting and roosting cavities. The red-cockaded woodpecker drills small holes in the bark and sapwood of a pine's living trunk, all around the cavity entrance. The exuding resin flows down the trunk, apparently repelling snakes and other predators and making a distinctive signpost in the pine forest. Egg-laying extends from late April to early June. These woodpeckers are cooperative breeders with helper birds aiding a mated pair with incubation, feeding, and brooding nestlings. Red-cockaded woodpeckers eat a variety of wood-boring insects as they spiral around or climb up the trunk. **THREATS:** The specific habitat requirements of the red-cockaded woodpecker have placed it in a losing conflict with current lumbering practices in southeastern Virginia. Old-age trees, usually incompatible with commercially managed pine plantations, are gradually being cut down. Fire, which once maintained the park-like stands preferred by the species, is rarely utilized as a management tool. Competition with other cavity nesters may be of considerable importance to the small Virginia population as habitat is lost to lumbering. **NOTES:** This woodpecker is an uncommon and local resident in open pine woodland from eastern Texas and Oklahoma eastward to southern Virginia and Florida. In Virginia, all red-cockaded woodpecker colonies exist on timber company properties on which old-growth pines are scarce and declining in numbers.

Abridged from "Red-Cockaded Woodpecker" by Ruth A. Beck, pp. 513–514 in *Virginia's Endangered Species*, coordinated by Karen Terwilliger, 1991 (Blacksburg, VA: The McDonald & Woodward Publishing Company).

Bewick's Wren

Thryomanes bewickii

ENDANGERED

DESCRIPTION: This is a medium-sized wren with an unstreaked brown back and conspicuous white eye-stripe. White spots on the tail show when the tail is spread or flicked sideways. Song resembles that of the song sparrow, usually 2 to 5 notes followed by a trill. Total length 4½ inches. DISTRIBUTION: Bewick's wren was formerly widely distributed in the mountains of Virginia, but it now breeds in only a few isolated localities in the higher mountains. HABITAT: This wren inhabits areas of sparse vegetation containing hawthorn, roses, fencerows, snags and downed logs, rock piles and outcrops. Around towns and farms it nests in outbuildings and cavities in old fence posts, conditions typical of the small farms and the spacious backyards of small towns of eastern North America in the 19th and early 20th centuries. LIFE HISTORY: Nesting probably begins soon after birds return from their wintering grounds in mid-April. The nest is loosely constructed of green mosses, sticks, dead leaves, cotton, hair, wool, and cast-off snake skins with an inner lining of feathers. An average clutch is 5 to 7 white eggs, irregularly dotted with browns, purples, lavender, and gray. Some pairs produce two or three broods a summer. Bewick's wren forages on the ground, on tree limbs and in the foliage, around outbuildings, in log piles, and in hedgerows. Insects compose 97 percent of the diet. THREATS: The reasons for the decline of Bewick's wren over the last 85 years remain unclear. The strongest correlation seems to be with land-use changes: the virtual disappearance of the small farm and changes in the size and structure of small towns, where Bewick's wren was formerly abundant. NOTES: Bewick's wren has been known to nest in almost any cavity, including tin cans, clothing hanging in buildings, baskets, and crevices in stone, brick, and tile walls. It has nearly vanished from much of its range east of the Mississippi River but continues to inhabit western and southwestern states. Bewick's wren winters in parts of its breeding range into Mexico, the Gulf Coast states, and Florida.

Abridged from "Bewick's Wren" by Curtis S. Adkisson, pp. 518–520 in *Virginia's Endangered Species*, coordinated by Karen Terwilliger, 1991 (Blacksburg, VA: The McDonald & Woodward Publishing Company).

Loggerhead Shrike
Lanius ludovicianus

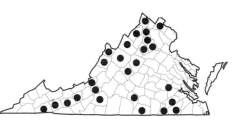

THREATENED

DESCRIPTION: The adult is gray above, white below, and has a conspicuous black "mask" through the eyes. It has a large head and a heavy, hooked bill. Total length is 7 inches. **DISTRIBUTION:** Confirmed breeding or potential breeding has occurred at one time or another in all five physiographic provinces of Virginia. Most recent records are from the mountain regions of the state, primarily the Shenandoah Valley and the outer Piedmont and Coastal Plain. Loggerhead shrikes also winter in the state. **HABITAT:** The loggerhead shrike occupies open country with scattered trees and shrubs. Typical breeding and wintering habitat in Virginia consists of closely grazed pastures with fencerows of shrubs and trees, which are used for nesting, perching, and roosting. Red cedars and hawthorns are frequently used for nesting. **LIFE HISTORY:** Egg-laying extends from early April to mid-June. Average clutch size is five eggs, spotted and colored dull white to light gray or buff. Some adults remain on the breeding grounds through winter, although during cold spells with snow cover they may move from hedgerows into woodlands. Invertebrates are the chief food throughout the year, but birds and small mammals are also taken. **THREATS:** Reasons for the recent decline of this species are incompletely known. Clearly habitat destruction as a result of clearing hedgerows, reforestation, and conversion of pasture to row-crops has taken its toll. Excessive winter kill may be a

problem, as may contamination by toxic chemicals. **NOTES:** Shrikes are known for their habit of impaling prey on thorns, barbed wire, or other sharp objects. The results of the Breeding Bird Surveys between 1984 and 1989 showed a significant decline of loggerhead shrikes in Virginia. Landowners need to become more aware of the value of hedgerow habitats to loggerhead shrikes and other species.

Abridged from "Loggerhead Shrike" by James D. Fraser, pp. 520–522 in *Virginia's Endangered Species*, coordinated by Karen Terwilliger, 1991 (Blacksburg, VA: The McDonald & Woodward Publishing Company).

Bachman's Sparrow

Aimophila aestivalis

THREATENED

DESCRIPTION: Bachman's sparrows are gray brown above, heavily streaked with chestnut or dark brown. They have a relatively large bill, flattish forehead, and long rounded tail. They are the only reddish brown sparrows in their habitat. The birds tend to be secretive but can be located by their distinct song, a clear whistle note followed by a trill on a different pitch. Total length is 5½ inches. **DISTRIBUTION:** Until 1950, this species was a scarce transient or summer resident in parts of the Piedmont and southwestern mountains. Recently active breeding sites have been found in Brunswick and Sussex counties. **HABITAT:** Present known breeding sites in Virginia are in abandoned fields and open pine plantations with abundant broom sedge. **LIFE HISTORY:** Nesting activities begin upon arrival from wintering grounds in March and April. Nests are built on the ground, partly concealed by grass tufts or low shrubs; typically they are domed with grasses. Three to five white eggs are laid by mid-May. Males continue to sing their song throughout the breeding season, even into August. Bachman's sparrow forages on the ground for beetles, weevils, bugs, grasshoppers, crickets, spiders, snails, and millipeds. Seeds of pines, grasses, sedges and wood sorrel are also important food items. **THREATS:** Cause for the general decline of Bachman's sparrow over its entire range are poorly understood. Habitat alteration or destruction is probably a major contributing factor. **NOTES:** Historically, Bachman's sparrow bred from Missouri eastward through Illinois to Pennsylvania and Maryland, southward to Oklahoma, Texas, the Gulf states, and Florida. It is rare to uncommon throughout most of its range and is extremely rare in Virginia.

Abridged from" Bachman's Sparrow" by Susan E. Ridd, pp. 522–523 in *Virginia's Endangered Species*, coordinated by Karen Terwilliger, 1991 (Blacksburg, VA: The McDonald & Woodward Publishing Company).

Henslow's Sparrow

Ammodramus henslowii

THREATENED

DESCRIPTION: Henslow's sparrow is a secretive bird with a large head, pale gray bill, and short tail. The head is olive with a yellowish crown stripe that becomes olive and streaked at the back of the crown. The brown back is streaked black, rust, and white; breast and sides are streaked with black; wings are chestnut. Henslow's sparrow usually sings from perches on low vegetation or from the ground while concealed in vegetation; it utters a short *tis-lick* or *flee-sic*. Total length is 4½ inches. **DISTRIBUTION:** Henslow's sparrow was formerly a transient and summer resident throughout most of Virginia and more common east of the Blue Ridge. Since the 1940s, it has experienced a severe decrease in numbers and is now rare to uncommon throughout state. **HABITAT:** Henslow's sparrow inhabits low, wet meadows and abandoned agricultural fields in early successional stages. Much of its time is spent on the ground among tall grass and brush. **LIFE HISTORY:** Henslow's sparrow arrives on its breeding grounds in April or May. It commonly nests in loose colonies. The nest, a deep cup of grasses or forbs lined with finer grasses and hair, is concealed on or near the ground. Three to five creamy or pale greenish white eggs heavily blotched with shades of brown and lavender are laid from May through August. Henslow's sparrow forages on the ground and eats a variety of insects, other invertebrates, and grass seeds. **THREATS:** The drainage of wetlands and intensive cultivation in the 20th century have reduced the amount of breeding habitat available to this species. **NOTES:** Henslow's sparrow breeds from eastern South Dakota south to central Kansas and Missouri, east to southern New England and the northern Piedmont of North Carolina. The winter range extends across the Coastal Plain from South Carolina to southern Texas.

Abridged from "Henslow's Sparrow" by Larry J. Brindza, pp. 525–526 in *Virginia's Endangered Species*, coordinated by Karen Terwilliger, 1991 (Blacksburg, VA: The McDonald & Woodward Publishing Company).

Mammals

In simplest terms, mammals are vertebrate animals that have hair on their bodies at some time during their lives and suckle their young with milk. There are about 4,000 species of living mammals in the world. Half of them are rodents, mostly mice and rats, and a quarter are bats. Most mammals are quadrupeds that creep, walk, run, or burrow on land. A few are arboreal climbers and leapers that spend much of their lives above the ground. Bats are the only mammals that truly fly (flying squirrels glide). A few dozen species are amphibious or aquatic.

Mammals representing all of these categories occur in Virginia, about 100 species in all (80 land mammals, 20 marine mammals). Fifty-five percent of Virginia's land mammals have nearly statewide distribution. Northern (boreal) species found primarily in the cooler mountainous counties make up 30 percent of the total. The other 15 percent are southern (austral) species of the warm lowlands east of the Blue Ridge. The occurrence of specialized habitats has a bearing on the distribution of some species. In the mountains such specialized habitats include caves, cliffs, talus, bogs, and spruce forests; beaches and extensive marshes and swamps are habitats peculiar to the eastern lowlands.

Virginia's pinniped fauna consists primarily of stray harbor seals, *Phoca vitulina*, most of which are young of the year. Adult harbor seals are very uncommon at our latitude. Gray seals, *Halichoerus grypus*, rare vagrants south of Massachusetts, occasionally may haul out on Virginia's beaches during severe winters. Two arctic or subarctic pinnipeds that have been reported as far south as Virginia are the hooded seal, *Cystophora cristata*, and the harp seal, *Pagophilus groenlandicus*. The presence of these species in temperate waters is extremely unusual and does not represent their normal distribution.

The only sea lion reported in the western North Atlantic is the California sea lion, *Zalophus californianus*. The number of sightings of these feral exotics, native in the North Pacific, has decreased since the Marine Mammals Protection Act of 1973 regulated the taking and keeping of marine mammals. California sea lions probably are not reproducing in the North Atlantic and sightings of them represent individuals that were released or have escaped captivity.

In the 1980s, knowledge of cetaceans of the coastal waters of Virginia increased at a remarkable rate, thanks largely to an increasingly efficient network for discovering and reporting strandings. In this decade, not only have hundreds of specimens of common species been examined, but four kinds of cetaceans have been recorded for the first time on the coast of Virginia.

Several other cetaceans have been found north and south of Virginia and eventually should be found on our coast. These include the sei whale (*Balaenoptera borealis*), blue whale (*Balaenoptera musculus*), short-snouted spinner dolphin (*Stenella clymene*), and killer whale (*Orcinus orca*).

From the 1800s to mid-1978, there were less than 100 cetacean records for Virginia. By mid-1989, the Marine Mammal Program of the Smithsonian Institution had accumulated over 550 records. In 1978, there were 95 cetacean records in Virginia, 69 of these were delphinids, among which were 45 bottle-nosed dolphin, *Tursiops truncatus*. As of mid-1989, approximately 444 of the cetacean records were delphinids, with 382 of these being bottle-nosed dolphins. Most were the result of the epizootic which occurred in 1987-1988 and the exact causes of which are still not understood.

It is estimated that roughly half of the inshore population of bottle-nosed dolphins died during the epizootic! Given the low fecundity of these animals, it may take the population more than 100 years to recover to the pre-1987 level, providing pressures on the surviving population remain at their current levels. With the dramatic increases in our utilization of our coastal resources as well as the omnipresent threat of pollution, one has to wonder how long it will actually take for the population to recover. In spite of the die-off, the bottle-nosed dolphin continues to be the most common inshore cetacean along the central and southern Atlantic coast.

Threats of long standing continue to plague endangered and rare mammals in Virginia. They include clearing and fragmentation of forests, removal of old growth, forest management without regard for nongame wildlife, intrusion of human beings into wilderness, marsh drainage, stream pollution, cave disturbance, and widespread development.

Some new threats loom large. Acid rain could well destroy the few remaining stands of spruce and fir. It may alter also the soil and water chemistry, and consequently, the litter and aquatic faunas important to shrews. The gypsy moth invasion could have a devastating impact on those species that are dependent on oak mast. Already the gypsy moth has been blamed for possibly contributing to the demise of woodrats in the northern United States. Loss of diversity in protected maturing forests is likely to eliminate the last snowshoe hares from Virginia. With the discovery of residual populations of the fox squirrel in southern Virginia, loss of mature oak-pine forest is now is recognized as a threat to the continued viability of that animal.

Many marine mammals are thought to be rare or endangered. Those that are, are rare or endangered throughout their entire ranges in the ocean at large, and all the states of the nation collectively must protect them.

Several kinds of inventories are needed. A statewide inventory of small mammals is long overdue. Such an inventory would provide information on distribution and material for study of systematics. In addition, special effort is need to map the real ranges of boreal relicts such as the water shrew, star-nosed mole, northern flying squirrel, and rock vole, and austral species such as the eastern big-eared bat, yellow bat, Seminole bat, marsh rabbit, and cotton mouse. Skulls of bobcats must be salvaged from trappers to form a transect of specimens between Dismal Swamp and the mountains. Then after taxonomic study, we can map the true distribution of *Felix rufus floridanus* in Virginia.

In dealing with rare or stressed species, population statistics such as numbers, geographic variation, demographics, and trends through time provide essential data for developing effective management plans. The Virginia Department of Game and Inland Fisheries should continue projects that monitor cave bats with carefully controlled searches of caves, the snowshoe hare with winter track counts, the fox squirrel on Assateague Island and the northern flying squirrel with nest box counts, and the fisher and mountain lion by recording sightings and investigating those that appear to be promising. Although river otter and bobcat are monitored routinely with hunting and trapping statistics, resolution could be improved with interviews of hunters, trappers, wardens, and biologists. Regular monitoring of known populations of the fox squirrel in Southside must be initiated as plans for management of this squirrel are developed.

Unfortunately, it is neither easy nor even practical to monitor some of the taxa for which we have concern. This is true of the shrews, star-nosed mole, eastern big-eared bat, eastern cottontail of the barrier islands, marsh rabbit, rock vole, cotton mouse, and least weasel.

The most opportune and prudent time to deal with special concerns, threats, and endangerment is before an organism comes into those categories. Not only is success more certain, it is also cheaper, easier, and quicker to prevent endangerment than it is to recover from it. A successsful endangered species program must be aware of the whole biota and not just its endangered parts. Inventories must be comprehensive and population monitoring must consider as much of the fauna and flora as resources will permit. For these reasons we recommend monitoring two species that are not listed as endangered or threatened — the mountain fox squirrel, *Sciurus niger vulpinus*, and the woodrat, *Neotoma floridana magister*.

Most of the remaining populations of the mountain fox squirrel are found in Virginia. They are stable or increasing in some areas, but they may be declining in northern and far southwestern Virginia. This taxon needs to be monitored periodically, perhaps at five year intervals, by means of hunting statistics and interviews with hunters, wardens, and biologists, to insure that declines can be identified and treated before they become serious.

In view of the plight of the woodrat north and west of Virginia, it would be prudent to monitor annually cliffs and ledges throughout Virginia known to be inhabited by this animal It is only necessary to record the location and quantity of fresh sign such as feces and feeding detritus.

In-depth studies of demography, natural history and ecology are needed at least for species that are endangered , threatened, of special concern, or that might come into these categories in the foreseeable future. There is an urgent need for study of the natural history of the woodrat.

Three animals require protective intervention. The status of the fox squirrel in northern Virginia needs review and, if justified, the ban on hunting this species east of the Blue Ridge should be extended to include northern parts of the state. To protect otter, beaver trapping should be banned in otter streams west of the Blue Ridge. In any case, western Virginia would be an ideal area for experimentation with means of separating beaver and otter trapping. Southwestern Virginia neesa a black bear sanctuary the equivalent of Shenadoah National Park and Great

Dismal Swamp National Wildlife Refuge. Genetic variability of bears in Dismal Swamp must be preserved by transplanting bears from other parts of the southeastern Untied States and not western Virginia.

Where areas are small, as they usually are in critical habitats, protection is not enough. Habitat must be actively managed with target species of flora and fauna in mind. Often what is good for one species is bad for another, so management strategies must be flexible and varied. Habitat can be created, such as by thinning the forest canopy or removing trees where laurel, rhododendron, and blackberry still exist, thus benefiting snowshoe hares (the hare would also benefit from planting cover plants such as red spruce) and by clearing undergrowth from oak-pine forest to benefit the fox squirrel. Roosts may be critical elements of the environment. Big-eared bats in Dismal Swamp could be encouraged with the construction of roost houses and cisterns. Appropriate nest boxes will attract flying squirrels and fox squirrels. It may be possible to create rocky labyrinths for the rock vole. After habitat has been improved or created and is well established, it may be necessary to transplant appropriate mammals to occupy new habitat. Isolated pockets of appropriate habitat may already exist for boreal taxa, no longer occupied by such species as water shrew and rock vole. Populations of these taxa might be reestablished.

The most critical habitats in Virginia are bat caves and boreal islands with red spruce. In Southside and on the Eastern Shore, the critical habitat is open oak-pine forest suitable for the fox squirrel. Also in Southside, undisturbed freshwater marshes are essential for the marsh rabbit, and early successional stages on swamp edges are necessary for southeastern shrews. Land acquisition for conservation purposes should have high priority in critical habitats; purchases, donations, swaps, and use of easements, as appropriate, may be necessary.

Because cavern-dwelling bats occupy only a small portion of available cave habitat, they are particularly vulnerable to disturbance. Low reproductive rate (only one or two young per year) and longevity (five to seven years on average) make bats more imperiled by negative population pressures than is the case with short-lived animals which have larger litter size or several litters per year. Further, disturbance of hibernating roosts takes its tool on bats. "A single human entry into a hibernation cave can cause thousands of bats to awaken prematurely and expend anywhere from 10 to 30 days worth of stored fat reserve, a supply which must last until spring in order to avoid starvation."[1]

Animals cannot be properly protected and managed until they have been studied and understood. Scientific endeavors usually do disturb wild populations, thus other human activity, such as recreational caving which offers no benefits to the survival of bats, should be minimized. "Caving seasons" — times when it is safe for people to enter bat caves (i.e., when bats are not present) — should be established. Most Virginia caves can be entered during the summer because the vast majority of them are not used by bats from mid-May through mid- to late August.

Measures should be taken to protect roosting sites. Properly designed gates on cave entrances are the most secure and afford the best protection but must be designed with the inhabitants of the cave in mind. Other protective measures include limiting the use of insecticides and preventing destruction of foraging habitat.

[1] From Tuttle, M.D. 1988. *The Importance of Bats.* Austin:Bat Conservation International.

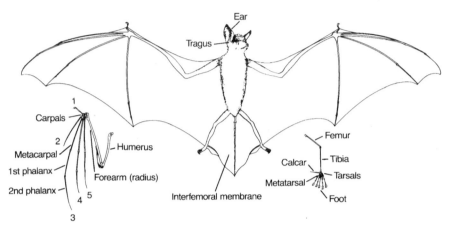

Figure 10. A generalized bat showing selected morphological characters.

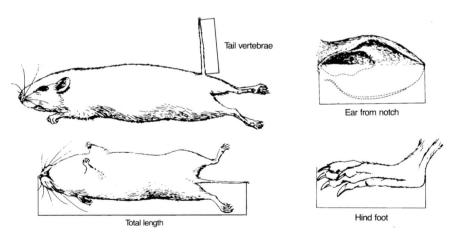

Figure 11. A generalized rodent showing selected morphological measurements.

Southeastern Shrew

Sorex longirostris fisheri

DESCRIPTION: The southeastern shrew can be recognized as a shrew by its mouse-like form, delicate feet, conical snout, tiny eyes, and small ears almost hidden in fur. This long-tailed shrew is one of the smallest ($^1/_{10}$ to $^2/_{10}$ ounce) mammals; within its range only the pygmy shrew, *Sorex hoyi*, is smaller. The southeastern shrew is dark brown, slightly paler beneath, and has buffy feet and an indistinctly bicolored tail. Total length is about 3¾ inches. It is about 20 percent larger than its upland counterpart *Sorex longirostris longirostris* and usually is duller in color above and more tinged with drab or wood brown below. **DISTRIBUTION:** This subspecies of the southeastern shrew is believed to occur only in the historical Dismal Swamp. Much of the present known range is within the boundaries of the Great Dismal Swamp National Wildlife Refuge. **HABITAT:** This subspecies of the southeastern shrew is most abundant in early to mid-successional forest habitats with a dense understory, moderate leaf litter, and moist organic soils. **LIFE HISTORY:** Very little is known about reproduction in this subspecies. Nest are constructed of dry leaves and other material placed under logs and hollow stumps. Most litters are believed to be born during June and August. Even less is known about activity periods, feeding habits, and behavior. Isopods (sowbugs and pillbugs) and amphipods (small shrimp-like animals), abundant in leaf litter, may be important foods. **THREATS:** Drainage of the Dismal Swamp and surrounding areas and drying of the Swamp permits invasion by the upland subspecies. Interbreeding of the two subspecies could bring about genetic extinction of *Sorex longirostris fisheri*. **NOTES:** Other common names for this shrew are Fisher's shrew, Dismal Swamp shrew and Dismal Swamp long-tailed shrew. The preservation of the forested wetlands of the Dismal Swamp is crucial to the survival of this subspecies. The extensive network of ditches constructed in the Swamp in the past two centuries and the overall loss of water in parts of the Swamp pose a direct threat not only to *Sorex longirostris fisheri* but to the entire wetland ecosystem.

Abridged from "Southeastern Shrew" by Robert K. Rose and Thomas M. Padgett, pp. 562–564 in *Virginia's Endangered Species*, coordinated by Karen Terwilliger, 1991 (Blacksburg, VA: The McDonald & Woodward Publishing Company).

Water Shrew

Sorex palustris punctulatus

ENDANGERED

DESCRIPTION: The water shrew is handsome with its glossy gray black back, silvery buff underparts, whitish hands and feet, and long, sharply bicolored tail. Toes and sides of hands and feet are fringed with stiff hairs to increase surface area. Enlarged feet with notably longer fringes of hair are highly effective paddles for swimming. This is the largest long-tailed shrew in eastern North America: total length is 6 inches, weight, ⁴/₁₀ to ⁵/₁₀ ounce. Although mouse-like in general appearance, the water shrew cannot be readily confused with any other local mammal. DISTRIBUTION: The subspecies *Sorex palustris punctulatus* occurs as a series of disjunct populations at high elevations in the southern Appalachians. Probably once widespread at higher elevations in Virginia, the water shrew has been found in recent times only along Little Back Creek in northwestern Bath County and at several sites in the Laurel Fork drainage in northwestern Highland County. HABITAT: Specimens from Bath County, Virginia, were taken in or near a small rocky stream in a narrow, steep-sided valley in a forest of beech, yellow birch, and sugar maple. Habitat at Highland County sites is similar except that forests include red spruce or Canadian hemlock. LIFE HISTORY: The water shrew is seldom found far from water. It obtains most of its food, predominantly immature stages of aquatic insects, in the water. It also eats the fungus *Endogone*, some plant material, and various animals including small fish. THREATS: Current threats are many: fragmentation of suitable habitat, leaving little opportunity for movement of shrews, even in relatively small areas; warming and siltation of headwater streams and ponds that result from logging, clearing for agriculture, and roadbuilding; acid rain and its effects on forests and waters that provide the shrew's habitat and food supply; and habitat loss and pesticide poisoning from spraying for gypsy moths. NOTES: The only locality in Virginia at which the rock vole, *Microtus chrotorrhinus*, has been found is in the Little Back Creek drainage, about one mile from the site where the water shrew was found. Populations of the water shrew in the southern Appalachians are relicts of the last Ice Age. Because of climatic warming and retreat of the water shrew's preferred boreal habitats to higher elevations, populations of this mammal must have been declining for thousands of years. Turn-of-the-century logging accelerated this demise.

Abridged from "Water Shrew" by John F. Pagels and Charles O. Handley, Jr., pp. 564–565 in *Virginia's Endangered Species*, coordinated by Karen Terwilliger, 1991 (Blacksburg, VA: The McDonald & Woodward Publishing Company).

Gray Myotis

Myotis grisescens

DESCRIPTION: The gray myotis is a medium-sized, mouse-eared bat. Its fur is short, dusky or reddish brown above; paler, whitish below. Dorsal hairs are similar in color from base to tip and not sharply defined blackish at the base, as they are in other small-eared eastern American bats. Total length, 3½ inches. The gray myotis is the only bat of eastern North America with almost unicolor dorsal hairs and wing attached at the metatarsus. DISTRIBUTION: The gray myotis has been found during spring, summer, and fall in the extreme southwestern part of Virginia in the drainage of the Tennessee River. HABITAT: The gray myotis lives in caves year-round. Males and females hibernate together in great aggregations during the winter in large, cold, complex caverns. Males and females migrate to separate roosts for the summer, with females forming large maternity colonies (up to thousands of individuals) in warm caves and males congregating in smaller numbers in other caves. Maternity roosts are usually located in caves through which large streams flow and which are usually close to rivers or reservoirs where the bats forage. LIFE HISTORY: Mating occurs in late fall. Young are born in late May and early June and are volant by early to mid-July. Only one young is produced per female each season. Gray myotis generally forage over water or in forested areas between the cave roost and water. Food consists of flying stages of aquatic insects, particularly mayflies. Females are the first to migrate, in early September, to winter hibernacula. They are followed by juveniles and males in mid-October. Some colonies migrate as far as 300 miles. Nearly all gray myotis hibernate in only five caves in the southeastern US. THREATS: Since this species aggregates in only a few caves, especially in the winter, it is especially vulnerable to local catastrophes and disturbance by human beings. Summer maternity colonies are particularly sensitive. Merely shining a light on clustered young will cause large numbers of them to drop to their deaths on the floor. Total numbers of the species have declined because of intrusion of cavers, destruction of foraging habitats, impoundment of waterways, commercialization of caves, and outright killings. NOTES: Other common names for the gray myotis include gray bat, Howell's bat, and cave bat. The gray myotis is listed federally and in Virginia as endangered.

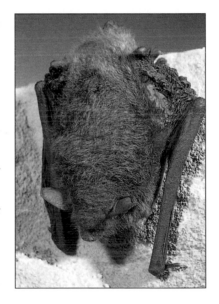

Abridged from "Gray Myotis: by Virginia M. Dalton and Charles O. Handley, Jr., pp. 567–569 in *Virginia's Endangered Species*, coordinated by Karen Terwilliger, 1991 (Blacksburg, VA: The McDonald & Woodward Publishing Company).

Social Myotis

Myotis sodalis

ENDANGERED

DESCRIPTION: The social myotis is a small brown bat with mouse-like ears and plain nose. Its pelage is smooth-lying and has a rather pastel pinkish brown hue. Total length, 3¼ inches. It can be recognized as a *Myotis* by its small ears and apparent small gap between canines and cheek teeth. It is distinguished from other species of *Myotis* in Virginia caves by the combination of keeled calcar, short toe hairs, and short ear. DISTRIBUTION: The social myotis hibernates in caves in the western part of Virginia, but it is seldom found in summer. HABITAT: Male and female social myotis congregate in the fall to hibernate in large caves and mine tunnels. Most males continue to use underground roosts in summer, but maternity colonies of up to 90 females form under loose bark of trees such as shagbark hickory, oaks, and maples. Bats emerge at night to feed on moths, flies, and other insects over tree-lined streams and upland woods. LIFE HISTORY: During autumn, social bats may migrate as much as 300 miles to caves where they spend the winter. By mid-October they have entered hibernation. On flat ceilings of large caves, they typically congregate in compact clusters containing 500 to 1,000 or more bats. Most clusters are found within a rather narrow zone within a cave, usually close to an entrance. The bats start emerging from hibernation in April. Females leave the hibernacula first. By mid-May very few bats remain in the caves. Mating occurs in the caves in the fall, but sperm are stored in the uterus until spring. Gestation period is unknown. The young are born toward the end of June or early in July. THREATS: The social myotis is vulnerable to natural disasters, such as flooding, as well as to pesticide poisoning and disturbance by human beings. This bat is easily disturbed by human activity — vandalism, spelunker traffic, cave commercialization, and scientific research. NOTES: Other common names for this bat include Indiana myotis, Indiana bat, social bat, and cluster bat. The social myotis is uncommon in Virginia. Numbers are declining here and throughout its range. The species is listed federally and in Virginia as endangered.

Abridged from "Social Myotis" by Virginia M. Dalton and Charles O. Handley, Jr., pp. 569–570 in *Virginia's Endangered Species*, coordinated by Karen Terwilliger, 1991 (The McDonald & Woodward Publishing Company).

Eastern Big-Eared Bat

Plecotus rafinesquii macrotis

ENDANGERED

DESCRIPTION: The eastern big-eared bat has enormous ears, more than twice the length of its head, and connected by a low band across the forehead; odd, mitten-shaped glandular masses on either side of the muzzle between nostril and eye; and elongate nostril openings. Its fur is long and rather shaggy, yellowish brown to reddish brown on the dorsum, white or whitish on the underparts, with sharply defined blackish hair bases throughout. Total length, 3½ to 4¼ inches. DISTRIBUTION: The eastern big-eared bat is known in Virginia only by a few occurrences in the southeastern part of the state. HABITAT: The eastern big-eared bat most often is found in houses, sometimes in hollow trees or behind loose bark, and rarely in culverts, caves, and mine tunnels. LIFE HISTORY: This species roosts singly, in small clusters, or in larger groups containing as many as 100 or more individuals. The species is nocturnal but not crepuscular. Food habits have not been studied. Mating apparently occurs in fall and winter; single young are born in May and June. Young bats can fly at three weeks of age and reach adult size in about four weeks. Longevity is 8 to 10 years. THREATS: Our ignorance of the ecology and natural history of the eastern big-eared bat makes it especially vulnerable to disturbance or destruction of its roosts and habitat. Unfavorable alteration of roosts or critical habitats, even by conservation agencies dedicated to protecting the bat, is a real possibility. Because these bats are sensitive to disturbance, are tame, and are so bizarre in appearance, they are subject to casual molestation. NOTES: Other common names for this bat include the eastern lump-nosed bat, eastern long-eared bat, Rafinesque's big-eared bat, and LeConte's big-eared bat. It is on the edge of its range in Virginia, so its occurrence here is tenuous at best.

Abridged from "Eastern Big-Eared Bat" by Charles O. Handley, Jr. and Don Schwab, pp. 571–573 in *Virginia's Endangered Species*, coordinated by Karen Terwilliger, 1991 (Blacksburg, VA: The McDonald & Woodward Publishing Company).

Western Big-Eared Bat

Plecotus townsendii virginianus

ENDANGERED

DESCRIPTION: The western big-eared bat is a medium-sized bat with huge ears connected by a low band across the forehead, mitten-shaped glandular masses on the muzzle, and elongated nostril openings. Fur is long and somewhat lax; hairs of the back are dark brown at tip and pale brown at base; underparts are buffy or pale cinnamon brown, hairs gray-brown at base. Total length, 4 inches. DISTRIBUTION: The range of this subspecies is fragmented into several populations, the most extensive of which occupies a dozen or more caves in tributary valleys of the Potomac and James rivers. An isolated population is found in Tazewell and Bland counties. HABITAT: These bats live year-round in caves at elevations greater than 1,500 feet above sea level. Although a colony may occupy the same cave throughout the year, it may have one or more other roosts in other caves. Individuals will move from one roost to another at any season, even in cold weather. Winter roost sites are often near cave entrances or in passages where there is considerable air movement. LIFE HISTORY: Females congregate in nursery colonies in just a few caves in summer. They start arriving at maternity caves in early to mid-April. Having mated during fall and winter, they arrive pregnant. Ordinarily, each female pro-

duces a single offspring. Young are born around mid-June, and by the end of July they begin to leave the cave at night to forage. Moths make up more than 90 percent of their diet. Most bats have left the maternity cave by mid- to late September. Males may also be found in caves in summer, scattered singly and in larger numbers, perhaps in bachelor colonies. Females and males hibernate together during winter months; solitary individuals and small clusters are the rule in hibernacula. There are many unanswered questions about fall and spring movements. THREATS: Both winter and summer populations are extremely intolerant of disturbance. Several winter colonies have disappeared. NOTES: Other common names for this bat are western lump-nosed bat, mule-eared bat, and Townsend's big-eared bat. About 2,000 occur in Virginia. This subspecies is listed federally and in Virginia as endangered.

Abridged from "Western Big-Eared Bat" by Virginia M. Dalton and Charles O. Handley, Jr., pp. 573–575 in *Virginia's Endangered Species*, coordinated by Karen Terwilliger, 1991 (Blacksburg, VA: The McDonald & Woodward Publishing Company).

184

Snowshoe Hare

Lepus americanus virginianus

ENDANGERED

DESCRIPTION: The snowshoe hare is characterized by large hind feet and seasonal changes in coloration of pelage. The subspecies in Virginia is the largest and most brightly colored of the snowshoe hares. Summer pelage is rusty brown dorsally, snout dark brown to black, nostrils edged with white, chin and belly whitish, tail white above and grayish beneath, and ear tips black. Winter pelage is entirely white, except for black-tipped ears and a brownish wash on feet and ears. Total length, 21 inches. DISTRIBUTION: The snowshoe hare is known in Virginia from only three sites in northwestern Highland County. Reports from elsewhere in the Alleghenies and in the Blue Ridge are unverified. HABITAT: The snowshoe hare is indigenous to the boreal forest. In Virginia, it is found in areas now or formerly forested with red spruce. The presence of adequate understory cover is probably the most critical component of its habitat. In Virginia, mountain laurel provides most of the suitable cover. LIFE HISTORY: The snowshoe hare consumes a wide variety of plant material. Hares forage mostly at night, and are usually solitary except during the breeding season. Females can come into heat the day young are born. A female may produce as many as four litters per year, each with one to eight young. Snowshoe hares make little use of dens or burrows. During the day, they hide in dense cover and remain motionless in a "form". From the form, a hare can easily leap out to avoid a predator. THREATS: Loss of habitat due to even-age succession of forest is the primary threat to remaining populations of snowshoe hares in Virginia. The overstory canopy of forest that once had thick stands of mountain laurel has closed to the point that the understory is thinning out. Within the snowshoe hare's current range there are very few areas with adequate cover remaining. The discontinuous nature of remaining pockets of cover have increased the vulnerability of hares to predation. NOTES: Other common names for the snowshoe hare include snowshoe rabbit and varying hare. The snowshoe hare is classified in Virginia as a game species with a continuous closed season.

Abridged from "Snowshoe Hare" by Michael L. Fies, pp. 576–578 in *Virginia's Endangered Species*, coordinated by Karen Terwilliger, 1991 (Blacksburg, VA: The McDonald & Woodward Publishing Company).

Northern Flying Squirrel

Glaucomys sabrinus fuscus

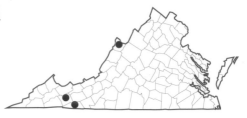

ENDANGERED

DESCRIPTION: The northern flying squirrel is a nocturnal squirrel with soft, silky fur and prominent, large black eyes. A loose fold of skin, which can be stretched out for gliding, extends from wrist to ankle. Total length, 10½ inches. Compared with the southern flying squirrel, *Glaucomys volans*, the northern species is larger and has gray-based belly fur (white-based in the southern species). DISTRIBUTION: In Virginia, this squirrel is known from only one site in Smyth County, three sites in Grayson County, and three sites in northwestern Highland County. HABITAT: In the southern Appalachians, the northern flying squirrel usually is associated with red spruce or Fraser fir and northern hardwood forest with moderate to thick evergreen understory. Most captures have been in moist forest with widely spaced, mature trees and an abundance of dead snags. Dependence on fungi may be a factor restricting the northern flying squirrel to moist sites at high elevations. It has not been captured in Virginia at elevations lower than 3,500 feet. LIFE HISTORY: The northern flying squirrel is omnivorous, subsisting mostly on fungi and lichens, but it also eats seeds, buds, fruits, staminate cones, catkins, tree sap, insects, and various animal material.

It occupies both tree cavities and outside nests and is often found aggregated in small family groups. Scanty data on reproduction indicate a single annual litter with two or three young born between March and May. Activity is negatively correlated with moonlight. THREATS: Populations of northern flying squirrel in the southern Appalachians probably have been declining since the end of the last Ice Age. Logging and clearing of spruce undoubtedly have accelerated this decline. Introduced insect pests, such as the balsam woolly aphid, and pollution by acid rain threaten to further reduce the habitat that remains. Alteration of habitat may have allowed the southern flying squirrel, which is more aggressive and capable of displacing the northern species, to spread into its range. NOTES: Probably fewer than 250 individuals remain in Virginia. The subspecies occurring in Virginia was federally listed as endangered on 1 July 1985.

Abridged from "Northern Flying Squirrel" by Michael L. Fies and John F. Pagels, pp. 583–584 in *Virginia's Endangered Species*, coordinated by Karen Terwilliger, 1991 (Blacksburg, VA: The McDonald & Woodward Publishing Company).

Fox Squirrel
Sciurus niger cinereus

ENDANGERED

DESCRIPTION: This is a large, heavy-bodied squirrel with an unusually full, fluffy tail. Upper parts of the body are dominantly whitish gray, occasionally with a buffy cast; underparts, hands, and feet white; snout and crown often white or whitish, or colored like adjacent parts of dorsum; cheeks whitish; ears whitish or buff; tail black and white with a white margin dorsally, and grayish with a submarginal black band and white margin ventrally. Total length, 24 inches; weight, 2 to 3 pounds. DISTRIBUTION: The subspecies *Sciurus niger cinereus* formerly occurred throughout the Eastern Shore. No remnant populations exist naturally in Virginia. However, successful restocking has occurred on Assateague Island, Chincoteague National Wildlife Refuge, and the fox squirrel has been reported on Chincoteague Island. A few survived restocking in Northampton County. HABITAT: The fox squirrel is found most often in open, park-like forests of mature loblolly pine and oak or in mixed stands of pine, beech, and sweetgum. Both upland and bottomland forests are occupied. Forests which contain a variety of nut and suitable seed bearing trees, contain over-age trees bearing hollows useful as den sites, and which have corn and bean fields nearby are especially attractive. LIFE HISTORY: Fox squirrels prefer dens in tree hollows. Leaf nests are most often situated in crotches of a tree trunk, in tangles of vines on a trunk, or toward the ends of larger branches, 35 to 50 feet above the ground. Fox squirrels of the Delmarva Peninsula feed largely on mature green pine cones. Mating occurs in late winter and early spring; most young are born in March or April. Average litter size is four. THREATS: The population on Assateague Island lives in overly mature loblolly pine forest on a barrier island in a region where pine-bark beetles are endemic.

Beetles, forest die-off, and severe storms all constitute threats to its habitat. NOTES: Other common names for this fox squirrel include Delmarva, Bryant, and peninsula fox squirrel. The fox squirrel is more cursorial and less agile than the gray squirrel. It is slower and more deliberate in its movements. When going from tree to tree, fox squirrels usually descend to the ground rather than leap from tree to tree. They are shy and wary and rather quiet. This subspecies was federally listed as endangered in 1967 and listed as endangered in Virginia in 1987.

Abridged from "Fox Squirrel" by Raymond D. Dueser and Charles O. Handley, Jr., pp. 585–587 in *Virginia's Endangered Species*, coordinated by Karen Terwilliger, 1991 (Blacksburg, VA: The McDonald & Woodward Publishing Company).

Rock Vole

Microtus chrotorrhinus carolinensis

ENDANGERED

DESCRIPTION: The rock vole is a small-eyed, short-tailed, shaggy haired rodent. Its long fur is overlain by coarse guard hairs that partially conceal its hairy ears. Its coloration above is brown, often with an orange cast; sides of snout bright orange rufous; underparts grayish white; feet grayish brown; tail indistinctly bicolored. Total length, 6¼ inches. The rock vole resembles the meadow vole, *Microtus pennsylvanicus*, but is somewhat more brightly colored. The nose spot of the meadow vole is always dull yellow, not orange, and smaller or even absent. Meadow voles and rock voles are not found in the same habitat. The rock vole does share its rocky habitat with the red-backed vole, *Clethrionomys gapperi*, which has slightly less shaggy fur, never has the sides of the snout orange, and has a back darker and differing from the flanks in coloration. **DISTRIBUTION:** This rodent has been found in Virginia only in the valley of Little Back Creek, Bath County. **HABITAT:** Typical habitat of the rock vole is cool, moist talus, or areas strewn with mossy boulders and logs. Streams, springs, or seeps are usually a few feet away. The voles are associated with red spruce, mixed forests of red spruce and northern hardwoods, and hemlock, oak, birch, and basswood forests, generally at elevations above 4,000 feet. **LIFE HISTORY:** This vole is active both day and night; most activity is confined to the labyrinths of its rocky habitat. The reproductive season extends from early March to mid-October. A female probably produces no more than two or three litters in a summer. The rock vole's diet is a mix of fruits, stems, foliage, and mushrooms. **THREATS:**

The cool, rocky, shady, moist habitat of the rock vole exists only sparingly in Virginia and is highly fragmented. It is easily destroyed by clearcutting. **NOTES:** The rock vole was common in Virginia at the end of the Pleistocene and probably was still present when the red spruce forests were cut in the late 1800s and early 1900s. Today it is known from only one locality, where its population must be relatively tiny. It is probably on the brink of extirpation in Virginia.

Abridged from "Rock Vole" by Charles O. Handley, Jr. and John F. Pagels, pp. 589–591 in *Virginia's Endangered Species*, coordinated by Karen Terwilliger, 1991 (Blacksburg, VA: The McDonald & Woodward Publishing Company).

Mountain Lion

Felis concolor couguar

ENDANGERED

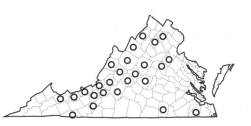

DESCRIPTION: The mountain lion is a large, long-tailed, plain-colored cat. Its fur is short and soft; upper parts tawny or some other shade of brown in summer, more grayish in winter; underparts whitish. Until they are about six months old, young mountain lions are paler, have a spotted coat, and ringed tail. Total length, 72 to 90 inches; weight, 80 to 200 pounds. Although such an animal should be unmistakable, views of it are usually fleeting, in poor light, and lacking perspective. Under such conditions it is possible to confuse dogs, bobcats, and even foxes with the big cat. More often tracks, up to 4 inches across and lacking claw marks, may be seen. DISTRIBUTION: Probably originally statewide in distribution, the mountain lion was thought to have been extirpated from the state in the late 19th century. After a lapse of about 75 years, reports of mountain lions began to come again from the mountain counties. While such reports are widespread and frequent, it is not known how many mountain lions there may be in Virginia, where they are, or whether they represent native *Felis concolor couguar* or introduced, exotic subspecies. HABITAT: The mountain lion possibly could live today in Virginia in extensive mountain hardwood forest or mixed forest with rock outcrops, ledges, and thickets of mountain laurel, rhododendron, and greenbrier. LIFE HISTORY: Males are solitary most of the year, but females may be accompanied by their young for up to two years after their birth. Mountain lions most often seek out large prey, preferably hoofed mammals such as deer and sometimes livestock. They also resort to a variety of smaller prey, including rabbits, squirrels, voles, beaver, smaller carnivores, birds, fish, and arthropods. The mountain lion stalks its prey and leaps upon it from the ground rather than from ambush in trees and rocks. It will hide uneaten portions of its kills for future meals, but it will not eat spoiled meat. THREATS: Human beings, with automobiles and guns as their weapons, are the only enemies of the mountain lion. NOTES: Other common names for the mountain lion include puma, panther, cougar, painter, catamount, and American lion. Once the most widespread mammal of the Western Hemisphere, with a range extending from ocean to ocean and from British Columbia to Patagonia, the mountain lion has long been persecuted and has been extirpated from most populated and agricultural areas. It is listed federally and in Virginia as endangered.

Abridged from "Mountain Lion" by Charles O. Handley, Jr., pp. 599–601 in *Virginia's Endangered Species*, coordinated by Karen Terwilliger, 1991 (Blacksburg, VA: The McDonald & Woodward Publishing Company).

Common Manatee

Trichechus manatus

ENDANGERED

DESCRIPTION: A massive mammal, the manatee has a broad head with large fleshy lips overhanging and hiding the lower jaw. It lacks hindlimbs, but has flipper-like forelimbs with rudimentary nails. Its large, flattened, spathulate tail is used for locomotion. The manatee's tough skin may be two inches thick. Adults are gray and often heavily scarred and covered with algae, barnacles, or other encrustations. Adults are 7 to 13 feet long and may weigh 1,300 pounds. There is no apparent sexual dimorphism. DISTRIBUTION: Manatees are restricted to warm waters (above 46°F) and usually are found south of Virginia and North Carolina. However, occasionally they travel north to Virginia waters. In recent years live animals have been sighted near the cities of Hampton, Portsmouth, and Hopewell as well as in the Chesapeake Bay and along the Atlantic coasts of Accomack and Northampton counties. HABITAT: Wholly aquatic, manatees inhabit warm, shallow coastal waters, estuaries, rivers, and canals. LIFE HISTORY: Sexual maturity is reached between four and nine years of age. Females usually breed once every two and a half to three years. One calf is born after a gestation period of about 11 months. No defined breeding season is known. Manatees feed almost exclusively on submerged aquatic vegetation. THREATS: A major threat to manatees, which are slow swimmers often found near centers of dense human population, is collision with power boat propellers. Habitat alteration from pollution, dredging, or filling as well as entanglement and drowning in fishing lines and nets also are significant mortality factors. NOTES: Manatees use their forelimbs for locomotion, for scratching, touching, and even embracing other manatees, for manipulating food, and for cleaning their mouths. Other common names include West Indian manatee, Florida manatee, and sea cow. The frequency of sightings of manatees in Virginia bays and rivers increased during the 1980s. The species is federally endangered and also protected by the Marine Mammal Protection Act.

Adapted from "Manatees *Trichechus manatus* Linnaeus, 1758; *Trichechus senegalensis* Link 1798 and *Trichechus inunguis* (Natterer, 1883)" by David K. Caldwell and Melba C. Caldwell, pp. 33–66 in *Handbook of Marine Mammals*, Vol. 3. *Sirenians and Baleen Whales*, edited by Sam H. Ridgway and Sir Richard Harrison, 1985 (NY: Academic Press).

Sei Whale

Balaenoptera borealis

ENDANGERED

DESCRIPTION: Sei whales can be distinguished from closely related species (blue, fin, Brydes, and minke whales) by their extremely large, dolphin-like dorsal fin. Like blue whales, sei whales are uniform dark blue-gray dorsally, whitish ventrally; frequently they have circular or oval scars that may resemble the blue whale's mottled appearance. The sei whale's baleen is typically gray-black with a paler fringe almost the texture of human hair. In the northern hemisphere, sei whales reach a maximum length of about 52 feet; females are slightly larger than males. DISTRIBUTION: The distribution of sei whales is poorly known. They are thought to migrate in summer to higher latitudes and in winter to temperate or equatorial waters. Very few sightings have actually been made in the western North Atlantic and most of these have been north of Virginia. However, sightings off Maryland, North Carolina, and Florida confirm that sei whales must pass through Virginia waters. HABITAT: This whale inhabits warm, deep water at the outer edge of the continental slope and further offshore in the open ocean. LIFE HISTORY: Sei whales attain sexual maturity between eight and ten years of age when they are 40 to 43 feet long. Calves are born during the March to September mating season and are typically 15 to 16 feet in length at birth. The calving interval is thought to be two years, with a gestation period of approximately one year and a lactation period of nine months. Unlike its close relatives which are categorized as gulp feeders or swallowers, sei whales are classed as skimmers like right, bow-head, and gray whales. Their preferred food is copepods; krill and fish are also consumed. THREATS: Sei whales have been observed towing fishing gear, and stranded sei whales have been found with net marks and other signs of obvious interaction with commercial fishing gear. NOTES: The sei whale is federally endangered and also protected under the Marine Mammal Protection Act.

Prepared by Charles W. Potter, Division of Mammals, National Museum of Natural History, Washington, DC.

Blue Whale

Balaenoptera musculus

DESCRIPTION: The largest animal in the world, an adult blue whale in the North Atlantic averages 80 to 85 feet long feet and may weigh 140 tons. Females are slightly larger than males. The back is a mottled bluish gray, the undersides of body and flippers are whitish, and the baleen is black. The head is broad and flat with a rounded snout. The dorsal fin is only about 13 inches long. The spout, which may reach 40 feet into the air, is narrow and vertical. **DISTRIBUTION:** Blue whales occur in all oceans. Strandings have been reported both north and south of Virginia; so this species must occur off Virginia as well. **HABITAT:** This is a species of the open ocean. It seems to prefer cold waters and to avoid warm waters. **LIFE HISTORY:** Blue whales are usually found alone or in groups of two or three individuals. They migrate to Arctic waters in spring to feed and spend the winter in temperate and possibly tropical waters. Blue whales feed almost exclusively on krill captured in the upper 300 feet of the sea. They apparently fast in winter. Mating and calving occur in late fall and winter. A single calf, about 25 feet long at birth, is born after an 11 month gestation period. Females reach sexual maturity at about age 5 years (when 68 to 75 feet long) and then produce one calf every two to three years. **THREATS:** Only a few hundred blue whales are believed to survive in the North Atlantic. The population is too small to insure long-term survival. Known mortality is due to whale-ship collisions and entrapment in polar ice. **NOTES:** The species most sought after by modern whalers armed with harpoon guns, the blue whale was hunted almost to extinction as late as the 1950s. The International Whaling Commission has protected the species since 1966. It is listed as federally endangered and is also protected under the Marine Mammal Protection Act.

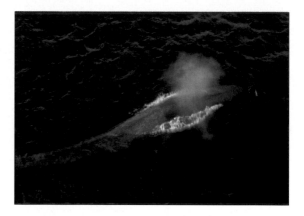

Adapted from "Blue Whale *Balaenoptera musculus* (Linnaeus, 1758)" by Pamela K. Yochem and Stephen Leatherwood, pp. 193–240 in *Handbook of Marine Mammals*, Vol. 3. *Sirenians and Baleen Whales*, edited by Sam H. Ridgway and Sir Richard Harrison, 1985 (NY: Academic Press).

Fin Whale

Balaenoptera physalus

ENDANGERED

DESCRIPTION: Fin whales are second in size only to the blue whale. Adults reach a length of 75 feet and may weigh 50 tons. They display a prominent dorsal fin about 24 inches long when they surface. Fin whales are more contrastingly colored than other great whales. The right lower jaw is white or ivory, the left jaw is black. The back is black dorsally, with an irregular white wash on the right side; underparts are whitish. The vertical spout is about 20 feet high. **DISTRIBUTION:** In Virginia, fin whales are commonly seen in the fall just offshore as they migrate from northern feeding grounds to southern wintering grounds. During the spring (March–April) northward migration of mackerel, fin whales can be seen feeding and traveling throughout the mid-Atlantic region. Subadults are found with some regularity in the Chesapeake Bay. **HABITAT:** Fin whales seem to prefer the offshore habitat from roughly 50 fathoms to the outer edge of the continental shelf. However, subadults may be found in the confines of inshore waters such as Chesapeake Bay. **LIFE HISTORY:** Both sexes reach sexual maturity between ages six and ten years, when they are 57 to 61 feet long. Mating probably takes place in winter. Gestation is thought to be 10 to 12 months; calves, approximately 21 feet long and weighing two tons, are born during the winter. Krill and schooling fish comprise the fin whale's diet. Feeding takes place primarily during the summer in the higher latitudes. **THREATS:** The greatest threats to the fin whale in Virginia waters appear to be ship strikes and entanglement in commercial fishing gear. **NOTES:** Fin whales may be the fastest of marine mammals, but they apparently sleep at the surface at night and thus are apt to be hit by passing ships. Once one was brought into Norfolk draped across the bow of a freighter. Fin whales are federally endangered and also protected under the Marine Mammal Protection Act.

Prepared by Charles W. Potter, Division of Mammals, National Museum of Natural History, Washington, DC. Additional information from "Swimming Beyond Boundaries — The Uncertain Future of Virginia's Marine Mammals and Sea Turtles," by Sue Bruenderman and Karen Terwilliger, pp. 11–27 in *Virginia Wildlife*, January 1994.

Northern Right Whale

Eubalaena glacialis

ENDANGERED

DESCRIPTION: Right whales are easily distinguished from other great whales by the absence of a dorsal fin or dorsal ridge; narrow, arched rostrum; long baleen (up to 7 feet); barnacle-covered callosities on head; lack of grooves on throat; robust stature; and low bushy spouts forming a V. Coloration is all black, or black with irregular white patches on the underparts. Adults reach a length up to 60 feet and may weigh 100 tons. Females are slightly larger than males. Only the bowhead is similar in shape and size, but it lacks callosities and has longer baleen (up to 14 feet). **DISTRIBUTION:** Right whales pass by Virginia on migration between their summering grounds north of Cape Cod and their calving and wintering grounds off the coasts of Georgia and Florida and elsewhere south of Virginia. A single immature right whale, estimated to be about 30 feet long, was sighted off Sandbridge on 18 February 1984. **HABITAT:** In the western North Atlantic, the right whale is found in temperate coastal waters, often close to shore. **LIFE HISTORY:** Calves are 12 to 18 feet long at birth. Sexual maturity occurs between the ages of five and nine years. In the northern part of their range right whales have been observed skim feeding on the surface and below on copepods and other invertebrates. Reportedly they also feed on small schooling fish. **THREATS:** Ship strikes and fishing gear entanglement seem to be key to the dire status of the right whale today. Other possible reasons for the failure of the population to recover from past over-exploitation by the whaling industry include inbreeding, pollution, and loss of undisturbed calving grounds. **NOTE:** Northern right whales had been targets of whalers since whaling began in the North Atlantic some 800 years ago. They were named "right whale" because they were the right whale to hunt, the only species that did not sink when harpooned. Today only about 350 remain in the western North Atlantic. No noticeable increase in population size has occurred since the international cessation of right whale harvest in 1937. They are federally endangered and also protected under the Marine Mammal Protection Act.

Prepared by Charles W. Potter, Division of Mammals, National Museum of Natural History, Washington, DC.

Humpback Whale

Megaptera novaeangliae

ENDANGERED

DESCRIPTION: The humpback is a short, stout whale with exceptionally long (to 13½ feet) narrow flippers. Adults may reach 50 feet or more in length and weigh 29 tons. The flukes are irregular in outline on the posterior edge and deeply notched in the center. The body is mostly black with irregular white blotches on the throat, abdomen, and sides. The flippers are white, and baleen is brownish black or gray. The body is irregularly dotted with fleshy bumps. The spout expands as much as 20 feet into the air. DISTRIBUTION: Humpbacks usually are on feeding grounds between Cape Cod, Massachusetts, and Iceland in spring, summer, and fall. In late fall they migrate to the Bahamas and spend the winter in warmer Caribbean waters, where they calve and breed. In recent years, several juvenile humpbacks have wintered off the Virginia coast. HABITAT: Humpbacks occupy coastal waters on their northern feeding grounds but apparently enter deep ocean during migrations. LIFE HISTORY: The humpback whale reaches sexual maturity between four and five years of age and physical maturity some ten years later. A female commonly breeds once every two years. Breeding occurs in winter, and gestation lasts 10 months. Newborn calves are 12 to 15 feet long. Humpbacks lunge-feed upon krill and fish trapped in a ring of bubbles produced by the whales' striking the water with their flippers. They are also known to follow trawlers apparently to feed on fish scared into schools by the passing ships. Humpbacks fast on the calving grounds. THREATS: Humpbacks are vulnerable to entanglement in gill nets and other types of fishing gear. NOTES: The scientific name of the species refers to the long flippers and translates as "great wing of New England." This is the famous "singing" whale also known for its breaching, "bubble-net" feeding behavior, and the ease with which ships can approach it. Humpbacks were particularly vulnerable to modern whalers due to their large aggregations on the calving grounds and their coming inshore on northern feeding grounds. The International Whaling Commission prohibited all hunting of humpback whales in the North Atlantic in 1956. The western North Atlantic population is now estimated at 5,000 individuals. The species is federally endangered and also protected by the Marine Mammal Protection Act.

Adapted from "Humpback Whale *Megaptera novaeangliae* (Borowski, 1781)" by Howard E. Winn and Nancy E. Reichley, pp. 241–273 in *Handbook of Marine Mammals*, Vol. 3. *Sirenians and Baleen Whales*, edited by Sam H. Ridgway and Sir Richard Harrison, 1985 (NY: Academic Press).

Sperm Whale

Physeter macrocephalus

ENDANGERED

DESCRIPTION: This toothed whale is distinguished by its huge, barrel-shaped head comprising ¼ to ⅓ of total body length. The jaw is narrow and rod-shaped. In place of a dorsal fin there is a conspicuous hump. Flippers are broad and rounded; the flukes are triangular with a straight trailing edge. The body is a uniform dark gray; upper lips and lower jaw are white. Often there are white blotches on the belly and flanks. There is marked sexual dimorphism in size: males may reach a total length of 69 feet, while the smaller females rarely exceed 38 feet. The single blowhole is on the midline and near the tip of the snout. The spout angles to the left from the head. DISTRIBUTION: The sperm whale is found in all deep oceans, from the equator to the edge of the polar ice packs. Only mature males migrate into the highest latitudes; females and immatures seldom venture poleward of 40°. HABITAT: This is a species of the open ocean. In Virginia waters, sperm whales may be encountered offshore near the 100-fathom line. LIFE HISTORY: Sexual maturity occurs between 9 and 13 years of age. The breeding season extends from January through August in the northern hemisphere. Calving occurs from May to September, after a 14 to 15 month gestation period. A female produces a single calf once every three to six years. The newborn is about 13 feet long. Sperm whales feed at great depths throughout the year. They are known for their long (60 to 90 minute), deep dives. Maximum depths near 10,000 feet have been reported, although most are within 3,000 feet of the surface. Top carnivores of the oceanic ecosystem, sperm whales feed primarily on the largest species of squid. They also take medium-sized squid, octopus, skates, cod, and sharks. THREATS: In calm weather, sperm whales will swim leisurely at the surface and even remain motionless for long periods of time making them vulnerable to collision with ships. Dead sperm whales have also been reported entangled at great depth in intercontinental telephone cables. NOTES: The sperm whale was a prime target of the American high seas whaling industry, which began in 1712. It was harvested for the spermaceti or whale oil in its head which was used as lamp fuel and as a lubricant.

The sperm whale fishery declined toward the end of the 19th century as petroleum came to replace whale oil. However, new uses for the high quality sperm oil in the 1950s led to a resurgence in whaling, which peaked in 1964. The International Whaling Commission banned sperm whaling worldwide in 1985. The sperm whale is federally endangered and also protected under the Marine Mammal Protection Act.

Adapted from "Sperm whale *Physeter macrocephalus* Linnaeus, 1758" by Dale W. Rice, pp. 177–234 in *Handbook of Marine Mammals*, Vol. 4. *River Dolphins and the Larger Toothed Whales*, edited by Sam H. Ridgway and Sir Richard Harrison, 1985 (NY: Academic Press). Additional information from "Swimming Beyond Boundaries — The Uncertain Future of Virginia's Marine Mammals and Sea Turtles" by Sue Bruenderman and Karen Terwilliger, pp. 11–27 in *Virginia Wildlife*, January 1994.

Appendix I

Definitions of Virginia and Federal Legal Status and Candidate Categories for Endangered and Threatened Species

A. Virginia*

Endangered	Any species which is in danger of extinction throughout all or a significant portion of its range, other than a species of the class Insecta deemed to be a pest and whose protection under the provisions of the article (§3.1–1021) would present an overriding risk to the health or economic welfare of the Commonwealth
Threatened	Any species which likely to become an endangered species within the foreseeable future throughout all or a significant portion of its range
Protected	All wild animals under the jurisdiction of the Virginia Department of Game and Inland Fisheries, except as otherwise permitted.
Special Concern	Any species which is restricted in distribution, uncommon, ecological specialized or threatened by other imminent factors.
Candidate species	A species formally recommended by the Director of the Department of Conservation and Recreation or other reliable data sources in writing to and accepted by the Commissioner for presentation the Board of Agriculture and Consumer Services for listing under the Virginia Endangered Plant and Insect Act.

*Definitions are from the Code of Virginia §3.1–1029, §29.1–521, and §29.1–563; VR 325–01, §14.

B. Federal*

Endangered	Any species which is in danger of extinction throughout all or a significant portion of its range other than a species of the Class Insecta determined by the Secretary (of Interior) to constitute a pest whose protection under the provisions of this Act would present an overwhelming and overriding risk to man.
Threatened	Any species which is likely to become an endangered species within the foreseeable future throughout all or a significant portion of its range.

197

Category 1	Taxa for which substantial information exists to support proposal to list the taxon as endangered or threatened.
Category 2	Taxa for which information exists to support proposal to list the taxon as endangered or threatened, but for which conclusive data on biological vulnerability and threat are not currently available to support proposed rules.
Category 3	Taxa that were once being considered for listing as endangered or threatened, but are not currently receiving such consideration.
Subcategory 3A	Taxa for which persuasive evidence of extinction is available. If rediscovered, such taxa might warrant high priority for addition to the List of Endangered and Threatened Wildlife.
Subcategory 3B	Taxonomic names that, on the basis of current taxonomic understanding usually represented in published revisions and monographs, do not represent taxa meeting the legal definition of species in the Endangered Species Act. Future investigations could lead to reevaluation of the listing qualifications of such entities.
Subcategory 3C	Taxa that are now considered to be more abundant and/or widespread than previously thought. Should new information suggest that any such taxa is experiencing a numerical or distributional decline, or is under substantial threat, it may be considered for transfer to category 1 or 2.

*Definitions of "endangered" and "threatened" from Endangered Species Act of 1973, as amended through the 100th Congress. Definitions of candidate categories condensed from 50 CFR 17 as reported in the *Federal Register*, volume 54 (4: January 6, 1989), pp. 554–555.

Appendix II

Information Sources for Changes in Legal Status of Species

The current official legal status of any species may be obtained from the appropriate regulatory agency, as follows:

State Endangered and Threatened Species

Animals (except insects):

Virginia Department of Game and Inland Fisheries
Nongame and Endangered Wildlife Program
P. O. Box 11104
Richmond, Virginia 23230-1104

Plants and insects:

Virginia Department of Agriculture and Consumer Services
Office of Plant Protection
P. O. Box 1163
Richmond, Virginia 23219

Federal Endangered and Threatened Species

Plants and animals:

Contact either:

United States Fish and Wildlife Service
P. O. Box 480
White Marsh, Virginia 23183

or

United States Fish and Wildlife Service
1825B Virginia Street
Annapolis, Maryland 21401

Appendix III

Species Listed as Endangered or Threatened in Virginia

Common Name	Scientific Name	Legal Status*
Vascular Plants		
Variable sedge	*Carex polymorpha*	SE
Harper's fimbristylis	*Fimbristylis prepusilla*	SE
Northeastern bulrush	*Scirpus ancistrochaetus*	FE, SE
Swamp pink	*Helonias bullata*	FT, SE
Prairie white fringed orchid	*Habenaria leucophaea*	FT, SE
Small whorled pogonia	*Isotria medeoloides*	FE, SE
Virginia round-leaf birch	*Betula uber*	FE, SE
Buckleya	*Buckleya distichophylla*	SE
Nestronia	*Nestronia umbellula*	SE
Addison's leatherflower	*Clematis addisonii*	SE
Millboro leatherflower	*Clematis viticaulis*	SE
Shale barren rockcress	*Arabis serotina*	FE, SE
Small-anthered bittercress	*Cardamine micranthera*	FE
Virginia spiraea	*Spiraea virginiana*	FT, SE
Northern joint-vetch	*Aeschynomeme virginica*	FT, SE
Running glade clover	*Trifolium calcaricum*	SE
Long-stalked holly	*Ilex collina*	SE
Michaux's sumac	*Rhus michauxii*	
Peters Mountain mallow	*Iliamna corei*	SE
Ginseng	*Panax quinquefolium*	FT, ST
Mat-forming water-hyssop	*Bacopa innominata*	SE
Smooth coneflower	*Echinacea laevigata*	FE
Virginia sneezeweed	*Helenium virginicum*	SE
Sun-facing coneflower	*Rudbeckia heliopsidis*	SE
Arthropods		
Freshwater crustaceans		
Madison Cave amphipod	*Stygobromus stegerorum*	ST
Madison Cave isopod	*Antrolana lira*	FT
Lee County cave isopod	*Lirceus usdagalun*	FE
Millipeds		
Ellett Valley pseudotremid milliped	*Pseudotremia cavernarum*	ST
Laurel Creek xystodesmid milliped	*Sigmoria whiteheadi*	ST
Insects		
Cherokee clubtail	*Stenogomphurus consanguis*	ST
Nantahala belted skimmer	*Macromia margarita*	ST
Swamp skimmer	*Tetragoneuria spinosa*	ST
Northeastern beach tiger beetle	*Cicindela dorsalis dorsalis*	FT

Rare skipper	*Problema bulenta*	ST
Regal fritillary	*Speyeria idalia*	*ST*

Mollusks

Freshwater mussels

Spectaclecase	*Cumberlandia monodonta*	SE
Dwarf wedgemussel	*Alasmidonta heterodon*	FE
Brook floater	*Alasmidonta varicosa*	SER
Slippershell	*Alasmidonta viridis*	SE
Tennessee heelsplitter	*Lasmigona holstonia*	SE
Little-wing pearlymussel	*Pegias fabula*	FE
Elephant ear	*Elliptio crassidens*	SE
Shiny pigtoe	*Fusconaia cor*	FE
Fine-rayed pigtoe	*Fusconaia cuneolus*	FE
Atlantic pigtoe	*Fusconaia masoni*	ST
Cracking pearlymussel	*Hemistena lata*	FE
Slabside pearlymussel	*Lexingtonia dolabelloides*	ST
Sheepnose	*Plethobasus cyphyus*	ST
James spinymussel	*Pleurobema collina*	FE
Ohio pigtoe	*Pleurobema cordatum*	SE
Rough pigtoe	*Pleurobema plenum*	FE
Pink pigtoe	*Pleurobema rubrum*	SE
Rough rabbitsfoot	*Quadrula cylindrica strigillata*	ST
Cumberland monkeyface	*Quadrula intermedia*	FE
Pimpleback	*Quadrula pustulosa pustulosa*	ST
Appalachian monkeyface	*Quadrula sparsa*	FE
Fanshell	*Cyprogenia stegaria*	FE
Dromedary pearlymussel	*Dromus dromas*	FE
Cumberland combshell	*Epioblasma brevidens*	SE
Oyster mussel	*Epioblasma capsaeformis*	SE
Tan riffleshell	*Epioblasma florentina walkeri*	FE
Green blossom	*Epioblasma torulosa gubernaculum*	FE
Snuffbox	*Epioblasma triquetra*	SE
Birdwing pearlymussel	*Lemiox rimosus*	FE
Fragile papershell	*Leptodea fragilis*	ST
Black sandshell	*Ligumia recta*	ST
Purple lilliput	*Toxolasma lividus*	SE
Deertoe	*Trucilla truncata*	SE
Purple bean	*Villosa perpurpurea*	SE
Cumberland bean	*Villosa trabalis*	FE
Pink mucket	*Lampsilis abrupta*	FE

Freshwater and land snails

Brown supercoil	*Paravitrea septadens*	ST
Rubble coil	*Helicodiscus lirellus*	SE
Shaggy coil	*Helicodiscus diadema*	SE
Spiny riversnail	*Io fluvialis*	ST
Spirit supercoil	*Paravitrea hera*	SE

Unthanks Cave snail	*Holsingeria unthanksensis*	SE
Virginia fringed mountain snail	*Polygyriscus virginianus*	FE

Freshwater Fishes

Shortnose sturgeon	*Acipenser brevirostrum*	FE
Paddlefish	*Polyodon spathula*	ST
Turquoise shiner	*Cyprinella monacha*	FT
Steelcolor shiner	*Cyprinella whipplei*	ST
Slender chub	*Erimystax cahni*	FT
Whitemouth shiner	*Notropis alborus*	ST
Emerald shiner	*Notropis atherinoides*	ST
Tennessee dace	*Phoxinus tennesseensis*	SE
Yellowfin madtom	*Noturus flavipinnis*	FT
Orangefin madtom	*Noturus gilberti*	ST
Blackbanded sunfish	*Enneacanthus chaetodon*	SE
Western sand darter	*Ammocrypta clara*	ST
Sharphead darter	*Etheostoma acuticeps*	SE
Greenfin darter	*Etheostoma chlorobranchium*	ST
Carolina darter	*Etheostoma collis*	ST
Tippecanoe darter	*Etheostoma tippecanoe*	ST
Variegate darter	*Etheostoma variatum*	SE
Duskytail darter	*Etheostoma sp.*	SE
Longhead darter	*Percina macrocephala*	ST
Roanoke logperch	*Percina rex*	FE

Amphibians
Frogs

Barking treefrog	*Hyla gratiosa*	ST

Salamanders

Mabee's salamander	*Ambystoma mabeei*	ST
Eastern tiger salmander	*Ambystoma tigrinum tigrinum*	SE
Shenandoah salamander	*Plethodon shenandoah*	FE

Reptiles
Turtles

Loggerhead sea turtle	*Caretta caretta*	FT
Atlantic green sea turtle	*Chelonia mydas*	FT
Atlantic hawksbill sea turtle	*Eretmochelys imbricata*	FE
Kemp's Ridley sea turtle	*Lepidochelys kempii*	FE
Leatherback sea turtle	*Dermochelys coriacea*	FE
Wood turtle	*Clemmys insculpta*	ST
Bog turtle	*Clemmys muhlenbergii*	SE
Eastern chicken turtle	*Deirochelys reticularia reticularia*	SE

Snakes

Canebrake rattlesnake	*Crotalus horridus atricaudatus*	SE

Lizards

Eastern glass lizard	*Ophisaurus ventralis*	ST

Birds

Bald eagle	*Haliaeetus leucocephalus*	FT
Peregrine falcon	*Falco peregrinus*	FE
Piping plover	*Charadrius melodus*	FT
Wilson's plover	*Charadrius wilsonia*	SE
Upland sandpiper	*Bartramia longicauda*	ST
Bewick's wren	*Thryomanes bewickii*	SE
Gull-billed tern	*Sterna nilotica*	ST
Roseate tern	*Sterna dougalii*	FE
Red-cockaded woodpecker	*Picoides borealis*	FE
Loggerhead shrike	*Lanius ludovicianus*	ST
Bachman's sparrow	*Aimophila aestivalis*	ST
Henslow's sparrow	*Ammodramus henslowii*	ST

Mammals

Terrestrial mammals

Southeastern shrew	*Sorex longirostris fisheri*	FT
Water shrew	*Sorex palustris punctulatus*	SE
Gray myotis	*Myotis grisescens*	FE
Social myotis	*Myotis sodalis*	FE
Eastern big-eared bat	*Plecotus rafinesquii macrotis*	SE
Western big-eared bat	*Plecotus townsendii virginianus*	FE
Snowshoe hare	*Lepus americanus virginianus*	SE
Northern flying squirrel	*Glaucomys sabrinus fuscus*	FE
Fox squirrel	*Sciurus niger cinereus*	FE
Rock vole	*Microtus chrotorrhinus carolinensis*	SE
Mountain lion	*Felis concolor couguar*	FE

Marine mammals

Common manatee	*Trichechus manatus*	FE
Sei whale	*Balaenoptera borealis*	FE
Blue whale	*Balaenoptera musculus*	FE
Fin whale	*Baleonoptera physalus*	FE
Northern right whale	*Eubalaena glacialis*	FE
Humpback whale	*Megaptera novaeangliae*	FE
Sperm whale	*Physeter macrocephalus*	FE

*FE - Federally endangered
FT - Federally threatened
SE - State endangered
ST - State threatened

Appendix IV

Species Recommended for Listing in Virginia

There are many species in Virginia which may be vulnerable to extinction but are not officially listed as threatened or endangered. The following species (with one exception) are those recommended for listing by the various taxonomic committees at the 1989 Symposium on Virginia's Endangered Species.

Common Name	Scientific Name	Recommended Status
Vascular Plants		
Lake quillwort	*Isoetes lacustris*	Endangered
Leather grapefern	*Botrychium multifidum*	Endangered
Slender lip fern	*Cheilanthes feei*	Endangered
Fraser fir	*Abies fraseri*	Endangered
Pondweed	*Potamogeton oakesianus*	Endangered
Wild chess	*Bromus kalmii*	Endangered
Plains muhly	*Muhlenbergia cuspidata*	Endangered
Bog bluegrass	*Poa paludigena*	Threatened
Awned sedge	*Carex atherodes*	Endangered
Barratt's sedge	*Carex barrattii*	Endangered
Epiphytic sedge	*Carex decomposita*	Endangered
Pale sedge	*Carex pallescens*	Endangered
Toothed sedge	*Cyperus dentatus*	Endangered
Granite flatsedge	*Cyperus granitophilus*	Endangered
Black-fruited spikerush	*Eleocharis melanocarpa*	Threatened
Robbin's spikerush	*Eleocharis robbinsii*	Endangered
Torrey's bulrush	*Scirpus torreyi*	Endangered
White buttons	*Eriocaulon septangulare*	Endangered
Pine barren rush	*Juncus abortivus*	Endangered
New Jersey rush	*Juncus caesariensis*	Endangered
Rush	*Juncus trifidus*	Endangered
White mandarin	*Streptopus amplexifolius*	Endangered
Bog rose	*Arethusa bulbosa*	Endangered
Showy lady's-slipper	*Cypripedium reginae*	Endangered
White fringed orchid	*Habenaria blephariglottis*	Endangered
Small's purslane	*Portulaca smallii*	Endangered
Virginia nailwort	*Paronychia virginica*	Endangered
Pursch's campion	*Silene ovata*	Endangered
Berlandier's anemone	*Anemone berlandieri*	Endangered
Western wall-flower	*Erysimum capitatum*	Endangered
Trumpets	*Sarracena flava*	Endangered
Small's stonecrop	*Diamorpha smallii*	Endangered
Large-leaved grass-of-parnassus	*Parnassia grandifolia*	Threatened

Dwarf saxifrage	*Saxifraga caroliniana*	Threatened
Sullivantia	*Sullivantia sullivantii*	Threatened
Bent milkvetch	*Astragalus distortus*	Endangered
Cooper's milkvetch	*Astragalus neglectus*	Endangered
Kate's Mountain clover	*Trifolium virginicum*	Threatened
Glade spurge	*Euphorbia purpurea*	Threatened
Cliff green	*Paxistima canbyi*	Threatened
Alder-leaved buckthorn	*Rhamnus alnifolia*	Threatened
Carolina lilaeopsis	*Lilaeopsis carolinensis*	Threatened
Bearberry	*Arctostaphylos uva-ursi*	Endangered
Pyxie moss	*Pyxidanthera barbulata*	Endangered
Fringed gentian	*Gentiana crinita*	Endangered
Buckbean	*Menyanthes trifoliata*	Endangered
Whorled horse-balm	*Collinsonia verticillata*	Endangered
Arkansas calamint	*Satureja arkansana*	Endangered
Eared tomanthera	*Tomanthera auriculata*	Endangered
Tall blazing star	*Liatris aspera*	Endangered

Insects

Mayflies

Benfields bearded small minnow mayfly	*Barbaetis benfieldi*	Endangered
Johnson's pronggill mayfly	*Leptophlebia johnsoni*	Endangered
Spieths great speckled olive	*Siphloplecton costalense*	Endangered

Dragonflies

Piedmont flaretail	*Gomphurus septima*	Endangered
Painted clubtail	*Gomphus apomyius*	Threatened
Friendly clubtail	*Gomphus parvidens*	Threatened
Blue faced clubtail	*Gomphus viridifrons*	Threatened
Speckled trout clubtail	*Lanthus parvulus*	Threatened
Minute dragontail	*Ophionuroides howei*	Endangered
Alleghany snaketail	*Ophionurus alleghaniensis*	Endangered
Piedmont snaketail	*Ophionurus incurvatus*	Endangered
Sapphire darner	*Aeschna mutata*	Threatened
Copper emerald	*Somatochlora georgiana*	Threatened

Stoneflies

Simmons slender winter stonefly	*Allocapnia simmonsi*	Endangered
Hallas broadback spring stonefly	*Prostoia hallasi*	Endangered
Nelsons early black stonefly	*Taeniopteryx nelsoni*	Endangered
Williams rare winter stonefly	*Megaleuctra williamsae*	Threatened
Lobed roach-like stonefly	*Tallaperla lobata*	Threatened
Flints common stonefly	*Aconeuria flinti*	Endangered
Beartown perlodid stonefly	*Isoperla major*	Threatened
Little Kanawha perlodid stonefly	*Diploperla kanawholensis*	Endangered

True bugs

Dismal Swamp green stink bug	*Chlorochroa dismalia*	Threatened
Opuntia squash bug	*Chelinidea vittiger*	Threatened
Sandpit alydid bug	*Stachyocnemus apicalis*	Threatened
Virginia Piedmont corixid bug	*Sigara depressa*	Threatened

Tiger beetles

Orange-bellied tiger beetle	*Cicindela abdominalis*	Threatened
Spectral tiger beetle	*Cicindela lepida*	Endangered

Butterflies and moths

Dukes skipper	*Euphyes dukesi*	Threatened
Appalachian grizzled skipper	*Pyrgus wyandot*	Threatened
Pink-edged sulphur	*Colias interior*	Threatened
Tawny crescent	*Phyciodes batesi*	Endangered
Herodias underwing moth	*Catocala herodias gerhardi*	Threatened

Fishes

Candy darter[1]	*Etheostoma osburni*	Threatened

Birds

Brown pelican	*Pelecanus occidentalis*	Threatened
Yellow-crowned night-heron	*Nycticorax violaceus*	Threatened
Northern harrier	*Circus cyaneus*	Endangered
Least tern	*Sterna antillarum*	Threatened
Sedge wren	*Cistothorus patensis*	Endangered
Swainson's warbler	*Limnothlypis swainsonii*	Threatened

Mammals

Fox squirrel	*Sciurus niger niger*	Endangered
Fisher	*Martes pennanti pennanti*	Endangered

[1]Noel Burkhead, personal communication, February 1994.

Appendix V

Species Believed Extinct or Extirpated in Virginia

Common Name	Scientific Name	Date of Last Verified Occurrence
Vascular Plants[1]		
Pine barren reedgrass	*Calamovilfa brevipilis*	1938
Pale beakrush	*Rhynchospora pallida*	1939
Seabeach pigweed	*Amaranthus pumilis*	?
Pond-spice	*Litsea aestivalis*	1805
Nuttall's micranthemum	*Micranthemum micranthemoides*	1941
Chaffseed	*Schwalbea americana*	1938
Heart-leaved plaintain	*Plantago cordata*	1915
Insect		
Dragonfly		
Alleghany snaketail[2]	*Ophionurus alleghaniensis*	?
Mollusks		
Freshwater mussels		
Round pigtoe	*Pleurobema coccineum*	
Rough pigtoe	*Pleurobema plenum*	
Acorn shell	*Epioblasma haysiana*	
Narrow catspaw	*Epioblasma lenior*	
Green blossom	*Epioblasma torulosa gubernaculum*	
Fragile papershell	*Leptodea fragilis*	
Cumberland bean	*Villosa trabalis*	
Pink mucket	*Lampsilis abrupta*	
Fishes[3]		
Harelip sucker	*Moxostoma lacerum*	1888
Shortnose sturgeon	*Acipenser brevirostrum*	1876
Thicklip chub	*Cyprinella labrosa*	1933
Trout-perch	*Percopsis omiscomaycus*	1911
Ashy darter	*Etheostoma cinereum*	1964
Northern logperch	*Percina caprodes semifasciata*	1938
Blackside darter	*Percina maculata*	1937
Birds		
Passenger pigeon	*Ectopistes migratorius*	
Carolina parakeet	*Conuropsis carolinensis*	
Mammals		
Beaver	*Castor canadensis*	1911[4,5]
Bison	*Bison bison bison*	1797[4]

Elk	*Cervus elaphus canadensis*	1855[4]
Elk	*Cervus elaphus nelsoni*	1974[4,6]
Fisher	*Martes pennanti*	1890[4,7]
Gray wolf	*Canis lupus lycaon*	1910[4]
Porcupine	*Erethizon dorsatum*	1837[4,8]

[1] An additional 33 species of vascular plants are today known only from historical records and may also be extirpated (see Porter, D., and T. Wieboldt, 1991. Table 22, p. 170, in "Vascular Plants," *Virginia's Endangered Species*, coordinated by Karen Terwilliger (Blacksburg: The McDonald & Woodward Publishing Company).

[2] Frank Louis Carle, personal communication, December 1993.

[3] Several other species of fish have been extirpated from entire drainage systems but not from the state as a whole.

[4] Handley, Charles O. 1979. "Recently Extinct or Extirpated [Mammals]," p. 584 in *Endangered and Threatened Plants and Animals of Virginia*, Donald W. Linzey, editor (Blacksburg: Virginia Tech).

[5] Beaver were restocked beginning in 1932 and have been successfully restored to Virginia's fauna.

[6] Elk from the Rocky Mountain subspecies were introduced to Virginia from 1917 to replace the extirpated native species. This subspecies also was extirpated.

[7] The fisher was restocked in West Virginia in 1969; sightings in Virginia are probably wanderers from the West Virginia population.

[8] Porcupines are occasionally sighted in Virginia, but it is believed that these individuals are vagrants from Pennsylvania or even hitchhikers on lumber trucks from the north.

Appendix VI

Credits for Photographs

American Society of Mammalogists: 190, 195
Tom Armour: 166
Roger W. Barbour: 155, 180
Robert E. Batie: 108
Peter W. Bergstrom: 165
D. Daniel Boone: 27
Dana Bradshaw: 164
Michael Britten: 30
Richard T. Bryant: 130, 135
Richard T. Bryant and W. C. Starnes: 123
Kurt A. Buhlmann: 153
Noel A. Burkhead: 122, 128
Noel A. Burkhead and Robert E. Jenkins: 117,
    119–121, 124–127, 129, 131, 132, 134, 136
Richard A. Byles, United States Fish and
    Wildlife Service: 148, 151
Ronald S. Caldwell: 112
Van Cotter: 42
Kenneth L. Crowell: 188
Steven M. Croy: 28, 31, 33, 36, 37, 48, 49
Jamie Doyle: 163
Sidney W. Dunkle: 59–61
Scott A. Eckert: 149, 152
Gary P. Fleming: 43
Thomas W. French: 179
Jessie M. Harris: 32, 35, 47
Richard L. Hoffman: 58, 63
C. Barry Knisley: 62
Richard K. LaVal: 181
Wayne Marion: 162
Christopher P. Marsh: 169

Diane McIntyre/Marine Mammal Images:
    191–194, 196
Joseph C. Mitchell: 144, 154, 156
Richard J. Neves and Steven Q. Croy: 72, 75–
    100, 102–107, 109, 110
Douglas W. Ogle: 39, 41
Paul A. Opler: 64
Johnny Randall: 26
Lynda Richardson: 55, 56, 145, 146
William N. Roston: 118, 133
William F. Ruska, Jr.: 29
A. E. Spreitzer: 73, 74, 101
W. Mark Swingle: 150
Merlin D. Tuttle, Bat Conservation
    International: 182–184
United States Fish and Wildlife Service: 189
Wayne Van Devender: 157
Vireo, Academy of Natural Sciences of
    Philadelphia: 167, 168, 170, 171, 173
Virginia Department of Game and Inland
    Fisheries: 185
Virginia Department of Agriculture and
    Consumer Services: 34, 44
Bryan Watts: 172
Thomas F. Wieboldt: 40, 46
Thomas F. and Alison B. Wieboldt: 38
Richard Webster: 187
Nancy M. Wells: 186
R. Harrison Wiegand: 45
George R. Zug: 147

Index of Common Names

Index of Scientific Names